The Science Of Faith

Scientific Persuasions
to enhance the
Spiritual and Religious Conviction

by

Vyner Mano

Chapter 0. Introduction

The author questioned his faith almost every day, faith in his definition is the belief in God. The author is not interested in comparing religion or discussing which religion is the better one or which religion makes more sense to be adopted. The author thinks the argument of religion is stupid, and religion at the current century probably has been corrupted by men's handling and taking over of the mantle ever since being revealed by God.

Having said that, the author is still a practising Catholic and he is also keen of science, every time he questioned his faith and how it can go wrong, he always found himself to be finding the answer through the law and manifestation of Science. The more he tried to combine science and religion the more he believed in God and his faith.

It is not unheard of that people of Science being non-believers of God, and they believe that the universe and everything that is known and unknown to mankind is governed and can be explained by Science.
For any faith devotees, they champion the idea of God as the almighty who created and is responsible for all the existences of things from the most minute thing to the beyond imaginable.

It is not entirely impossible to believe in both of the ideas without contradicting each other, instead this requires multiple levels of understanding- much like in the movie "Inception", within a dream comes another totally complete and complicated dream.

The same goes for science and faith. We can decipher the mysteries of faith and God as well as its creation through Science, and at the same time, we can make sense of many scientific occurrences with faith and God as the tool of reasoning. This is what the world has yet to come to an agreement though there is an awareness of how faith and science can be integrated together.

In this era of science and technology, information and system, exploration and connectivity, human beings are very much independent and it is easy to forget that above all of this there is a higher power that is governing us and the scientific law that is playing its role in order for us to behave as it is supposed to be.

When human is being stripped off of all of the technology and benefits and life is beginning to be put in a perspective, we will begin to question life and faith. Arguments on what is the purpose of all the happenings around us and some are beginning to doubt on the existence of a higher power- that is when God is interlinked: Everything needs to be explained with God fitting inside the picture.

In this book, the attempts to explain the Whys-of-life will be done using analogies and will be used throughout. Also, scientific knowledges will be discussed to make a case on why all of the phenomena and occurrences are revolving around God and the needs to have faith in the Higher Being.

Humans, as creation with the ability to think, will frequently conceive mind-boggling and sometimes life-changing thoughts which will be presented in this book. Also how much involved is God in our life in allowing happenings into our life will be discussed. Finally, it will be shown how scientific laws and rules are very much complimenting all the works that are already done by God and how it is slowly being manifested into our life all our lives.

The author is a practising Catholic and many examples may be based on Christian beliefs due to the author's familiarity with and confidence in sharing it in this book. However, the author grown up surrounded and meeting with many other religions such as Islam, Buddhism and Hinduism.

The author will provide examples that are relatable and comparable with the other beliefs in order for this book to be universally accepted or at least being given the chance to be read and understood to show that science and our faith is inter-linked and our faith is stronger when we put science into consideration.

Basic and simple science and physics theories will be used rather than mathematics or any computational calculations. The author and the book will be attempting to explain rather than providing proof of the concept of faith while relating it with components of the universe.

The science in this book will be oversimplified with a hope that those with less science-acumen will be able to grasp the scientific theory. At the same time, this book aspires to help the science community to connect the dots between science and faith.

It is also hoped that the science community will be able to get some help in explaining science to others and explaining faith to themselves.

This is The Science of Faith.

Chapter 1. Superposition

The very often word that is used to describe God is 'infinite'. His power is infinite, his knowledge goes all the way to infinity, however when human is trying to understand something, infinity doesn't really is the satisfying answer.

Saying the higher power equals infinity is actually a lazy way of explaining the concept of how God could have worked. In order for us to explain something that is as huge as God, we have to go into details, all the way into the minute scale of it. There could be times where we will question can God listen to your prayers and the others' at the same time?

How can God listens to multiple prayers and pay attention to all the details all at once, not taking into account the intervention needed to be done, on top of it deliberating whether a dream is in line with His own big plans without contradiction and mistakes? Of course, the easy way to answer the question is that God is all almighty and this is not impossible for him.

It is indeed very possible and can be done and explained. This phenomenon can be explained by an occurrence that is called 'Superposition'. Superposition is the ability to be in two positions simultaneously. Not only being at two different places at the same time but executing work at both places to produce different results but cumulatively resulting in a better and ultimate good outcome.

In a different way of saying, God has the superposition ability and it enables Him to work in all of our lives simultaneously and continuously without any problem, or at least the possibility of God leaving us out due to being occupied with some other work and projects is not happening. In order to explain how superposition is possible, a simple analogy needs to be provided in order for us to grasp the idea and go into detail of it.

Imagine there is a moving car, when looking at the bigger picture the car is governed by a lot of rules and laws such as forces of gravity exerted on it, the drag of the wind influencing it, the weight of the car is giving a say on it, aerodynamics of the car is playing it roles on the car and so on.

All of this is only looking at the exterior and the outside of the car. However, when focusing more into the details of the car especially in the interior and smaller scale, the car is governed by a lot more stuff and factors, making the car as it is.

When looking deeper into the car, there are the frames of the car the give the support to the car and serve as the main structural part of the car. The interior part of the car also has a non-leathered type of seats and cushion to absorb heat and reduces the heat on the surface of the seat. The car also has air-conditioning ability to provide with cool air without the need of the outside condition to achieve the desired temperature. Also, the car could have the radio playing songs to entertain the drivers and the passengers of the car.

The examples given about the interior of the cars are all independent parts that work differently but simultaneously and interlinking to serve the user of the car. All the parts are not serving the same purpose but as a whole, they are functioning as one complete car. Based on the analogy given, it is a simple concept that things can be done separately and simultaneously without a problem while serving on a bigger purpose.

However, superposition is better explained using the scientific terms.

Imagine there is a supercomputer, and that supercomputer is processing data and info in binary terms or Bits- meaning 1010, 1 for yes and 0 for no- every data is unique and it involves 1 and 0, and when there is a lot of it the binary number will keep chaining between 0 and 1. Taking into account the demand of the data to be processed, the speed needed to get it done must be extremely high, so the binary form could be 1 and almost instantaneously it could be at 0. When the speed is absurdly high in order for the data to be able to be processed and not crash, the binary terms need to be able to be a 1 and 0 at the same time.

This is called quantum bits or Qubits.

There would be thousands or millions of individual Qubits working at superposition in order to process and finish a bigger work and load. All of it is working individually and unique of each other but it is still governed and controlled by the big supercomputers.

To further explore how does superposition works, there is a need to explain what quantum is. Imagine there is an apple, apple when looking at the exterior, are governed by physics like mass, the volume of the density of the apple, and also external forces like gravity, and velocity of the wind.

The apple usually can go up when being thrown by human or any other machine and will go down due to the forces of gravity, it can also bounce off the ground when hitting into the ground. By zooming into the apple, apple is made of millions of atomic particles and at this sub-atomic level of size, how it reacts and works are governed by a smaller law called quantum physics. It does not longer being governed by common physics as we know it.

At a quantum level, or at sub-atomic phase, the particles can act either as a particle -imagine it as a sphere that can go up, down, either sideways and also it can spin. The particles can also act as a wave, moving up and down with a certain amplitude and wavelength.

At the quantum level, two particles can send messages to each other instantaneously although they are at a very vast distance to each other. Last but not least, a particle can be at two places at a time. These are called superposition ability.

Going back to the supercomputer, the processing computer is the law that governs all these Qubits, and the Qubits are being used of this ability to perform complicated jobs, at a high speed, simultaneously, with the other hundreds and thousands of other Qubits.

Having said that, all of this is possible since the whole system is huge enough, and within the system, it comprises a huge number of minuscule object to exhibit quantum physics behaviour.

Quantum physics, however, does not apply to all and for sure not at the traditional physics environment. One example of when superposition theory is not applicable: a pendulum swing. When you swing the pendulum, it will only be in one place at a time, no matter how fast it swings or how strong you pushes the pendulum before it swings, it can only be at one position at a time.

Meaning to say when you put your finger on the swinging path, it will hit the pendulum unless you have calculated the exact time when the pendulum is not at where your finger is. This is because the pendulum is governed by the normal physics.

Try to shift your attention to the second example: the water in the sea. The sea is huge and covers most parts of the world. But within the sea, at the quantum level it is made up thousands and millions of H2O. So it can be somewhere in America, and at the same time, it is where you are floating now, somewhere in the sea of Asia. The sea is great enough to possess this kind of character but at the end of the day, it governs each water molecules independently even though the molecules are so small compared to the whole sea. No matter how small is the water molecules, the sea knows what is happening in the overall system.

By now, the concept of Superposition has been explained and being given of analogy on how it works and what it is about. It is time to apply this idea to the omnipresence and omniscient of God, on how the ability is possible while getting logic into the equation. This is important in order for us to be convinced to put our trust that the higher power is in control and have the power and ability to listen to what we need, answer our prayers or to supply us with other solution.

The world has around 7 billion people and it is a lot of numbers for one person to handle and to cater for. But when we look at the world, the world is like an apple and when looking into the world, there are billions of people, just like the sub-atomic particles that made an apple. The whole population is like particles that are governed by quantum law, and we humans are all governed by God's power.

There is a saying: God with us; that means God is with each and every one of us. Speaking in quantum terms, God is always is superposition state to be able to work and to be with all of us around the world at the same time. All of us are working uniquely and independently, just like Qubits that are working actively all the time.

Each of us is representing one Qubit, that is working on a highly different thing with the other Qubits. There will be some similarities between one Qubit with the others but there will be no two exactly the same job or work that is being done, just like how there is no two similar thumbprint. But all of us are working together and live together to achieve bigger goals and targets.

We all will encounter problems and difficulties every day and sometimes the problem is a little bit hard for us to handle ourselves. Some storms are a little bit harder to go through in which we could use some help. Most of us at some point will turn to God for strength and guidance, with the trust that help will be given.

Prayer itself is unique with people, some are praying just to be given signs and hope in order for them to know that they are doing the right thing; some, however, have been given a tougher task at hand and sometimes the only thing that is left or the last resort is praying for God to get things done.

There are 3 ways of how prayers are usually being prayed:

The first way is to give thanks for all that you have been blessed with, to appreciate and praise the Almighty for all the provision and for the ability to have a good life. Be thankful to be able to look back and see that life is good and beautiful.

The second way of praying is to seek forgiveness. Some are very devoted to God but also acknowledged that they are very much weak and highly flawed that making mistakes and doing offences is a repetitive cycle. Some are praying for redemption for they have been strayed and they would like to come back: for they have realised that they have been wrong in what they used to thought that they are right- just like the story of the prodigal son: who left the house with his share of his father's wealth, he then spent all the money living hedonistically and one day found out that he was left with nothing and had to tend to the swine. He then decided to come back and seek forgiveness from his father in a hope that he could get back to him.

The third way of praying would be praying for own intention, some wanted to have something to do with health, some desired financial stability or freedom, some are hoping for romance and relationship to be good in life, some are praying for their faith and belief which sometimes is shaky and full of doubts.

All of these prayers are of various types and when taking each personality into consideration, the prayers get more complicated and complex. Taking into account the mixes of the types of prayer, that would be at least trillions of a probability of how the prayers could be. To be able to handle all of these workloads and demands, the higher being should be possessing the superposition ability.

With each prayer too, different reaction and solution need to be considered and provided. Some prayers required compassion when help is requested and pleaded, some prayers need to be treated with a degree of anger and disagreement, some payers need to be responded with coveting and delight, some prayers required decision making as it will affect others and also the life of the ones who prayed.

For every action, there will be a reaction. Every prayer will result in consequences, either actively or passively. How God responded and carry out the way forward will be explained later in the book, but with the explanation of superposition, it is possible for God to respond and produce an action simultaneously and in accordance with the master plan and the intention pleaded.

It takes a lot of effort and power to be able to cater every demand and process the information, consequently producing action to heed with the call and at the same time contributing to the master plan for each individual and the whole system as well. Let's take a few steps back again and try to look this in a more complicated and detailed perspective.

For now, we will dive into the visual representation of superposition a little bit to be able to immerse into this ability and characteristic. The idea of Superposition can be visually explained using a sphere- inspired by the Bloch Sphere - imagine north pole and south pole are on the sphere. The North Pole represents 0 of the bits, and the South Pole is 1 of the bits. Qubits are like the straight line that is connected from the North Pole all the way to the South Pole, so the straight line is at the North Pole and at the same time is also at the South Pole, it is at the 0 and also at the 1.

So whenever we move the straight line, the other end of the line is at the other side of the sphere which is at the extreme end, although the straight line can be 'folded' and both ends are facing the same location and spots on the sphere- representing the flexibility of the line and at the same time the ability of being at two places at once.

This is a good representation and visualisation of how God can be made available anywhere at the same time on Earth. The faith of the believer would be that the Divine Almighty is always there to listen and knows what he or she is experiencing and thinking.   This belief is powerful in ensuring that the believer can have someone to hang on to and to release all that is being kept inside.

There are also those who believe that there is God, but even though they are not praying or seldom pray, deep in their heart they acknowledged that everything that they have done or all the intention-based action that they committed are being observed and being monitored closely by God.   Emotionally speaking, this is good because it helps that person to not feel alone and he feels that everything that is being done in his life is with justifiable reason and can only be understood by someone that is walking in their shoes.   When a person is able to tell something about their life and the problem surrounding it, they are able to lift some of the burdens off their shoulders.

Theist usually boasts about how their beliefs are giving them the peace that the world cannot give, or they felt the disappearing emotional pain and tension thanks to the blessings by the divine power.   This is due to the recognition that the divinity power is close to them it provides them with the ability to cope and psychologically speaking, they are granted therapy and meditation.   This can only be achieved if God has the same ability as that of quantum superposition.

We can also demonstrate this in a virtual experiment that is inspired by the paper prepared by MIT students, with a little help of the experiment box, mirrors and a transparent box. The experiment box has three holes, one at the top side of the box, and adjacent to the hole, there is a hole at each side of the box.

The first hole at one side of the box is the entrance of a ray that is going to be emitted into the box, and there are two exits namely on top of the box and on the other side of the box. To identify the type of the ray, we have to see on which exit that the ray goes into, if it is a type "1", it will exit the top hole, whereas the other side of the hole will be where the ray exits if it is of type "2". To determine which side does the ray goes out of, there will be mirrors position at the travelling path of the ray exiting the box. From the mirror, it will reflect that ray and shine it into the transparent box.

When being shone with type "1" ray, the transparent box will turn white, and when being shone with type "2" ray and the box will turn into black. So when a ray is to exert both characteristic, simultaneously of type "1" and type, "2", the ray will exit on both holes located at the top and another side of the box opposite the entry. Both mirrors will be receiving the ray and they will reflect the ray into the transparent box.

The transparent box would receive the ray from both mirrors and registering that the ray is both type "1" and type "2", consequently the transparent box will have both white and black colour on its surface. This is one of the virtual experiment that can demonstrate how a thing can both be one type and at the same time be of another type while being at two different places at the same time to produce a different result.

The idea of superposition as the medium is inspired by the quantum computing that is simultaneously storing/ receiving information and processing calculation.

The big idea of quantum computing is to work under an algorithm that they can process data and preserve the correct info while recalculating and correcting any error while not corrupting the correct data. The Qubits that quantum computing is working is under superposition state to solve the problems at hand. Most problems are unique and highly complicated such as integer factorisation that involved breaking a number into products of smaller integers. By decomposing those numbers, more light can be shed to those areas in order to clear up the error that is existing.

When quantum computing is being explained, it can't be helped but to point out that there is a resemblance in how one can explain the omnipresence and superintendence of God. By acknowledging that the phenomenon of superposition exists and is working in everyday life, it can be considered that this is how God is able to work and perform the intervention in the living and the non-living things' life.

By being able to be in many places at one time and dictate things over a vast distance, it makes sense and logically to have the notion that the universe and objects within it are indeed being governed by the law which is the divine power of God.

But at the end of the day, one should be working hard to achieve everything in life, always carry out things as if you are the master of your own life and you are the determiner of how your life will turn out. At the same time, one will gain the many benefits psychologically and morally by praying to God. The science of praying will be explored more in the later chapters.

Saint Augustine once quoted, Work as if everything deepens on you, pray as if everything deepens on God. But again God's work is not usually a direct result, for example when you pray for God's help to keep your country safe, God will help you by giving you the courage and faith that you can do it, but the hard work and going out there defending and saving the country still needs to be done solely by our own effort with our own strong will.

Another example is when you wanted something in your life, you don't ask God to give it to you miraculously, but you need to get it yourself. Some of us sometimes resolved in cheating or shortcuts, but when you recognised your wrongdoings, you can always seek forgiveness from God. The concept of Divine intervention will be explored further in the later chapters. Superposition is a theory where an object can be of two states simultaneously at a time which can explain how a person can be at two places at one time.

However, there is another similar phenomenon on how one object's state can affect another object although they are separated at a very long distance, in other words, if object A is vibrating upwards, hence over a very large distance object B will be vibrating downward. This occurrence is called quantum entanglement. This phenomenon is trying to explain that every particle in the universe is so interconnected that the state at which they are in will be due to the state of the other related particles experiencing.

To make an analogy out of this is when the light is being shone into a crystal, some portion of the light leaves the crystal and goes to the left side of the room, then the rest of the light leaves spontaneously to the right side of the room. Interestingly, these two separated lights from the crystal each producing their own halos when hitting the walls of the room. Each halo is large enough that the two different halos are intersecting each other like a Venn Diagram.

That small area that the two halos are overlapping each other is an analogous representation of quantum entanglement: even though those two are two separate individuals, coming from two different locations but what happened to one halo is observable in another halo. They are entangled because in the end they are still one system as a whole and in the beginning, they come into existence from the same source. They are the part of the same creation which is Light.

Even though the two separated entities, having no communication to be happening between them, both of the observable light product somewhat know in what state they should be in and how should they react with each other. In reality, there should be no obvious entanglement or communication between the two objects, but for the sake of easy representation, imaginary halos are being used to show that they are 'entangled' enough to be behaving in a manner that there should be a reaction to any behaviour exerted by another object.

We know the light is a type of wave and waves have the property of exhibiting alternating shape. When observed, the light will give the impression that there are 'lines' that are queuing and moving vertically. Referring to the earlier analogy, if the light on the left is having waves that are seemingly moving upward, then the light on the right side wave would be perceivably moving downward.

This is a spontaneous reaction that shows that everything in the universe is linked and related to each other- no matter how far one is located from each other. This explains on how we experienced feelings and reaction if something is happening to another person, though we don't establish any communication to be updated of what another person is experiencing and the reaction that comes with it. Quantum entanglement will be further explored in the next chapter in relation to spacetime and time itself.

Quantum entanglement explains how God worked on human beings and the universe. When God intervenes, the universe and its living things would react and received the product. This is one of the ways how God work from a vast distance and how He can work with one person and simultaneously affecting other people as well.

Although, any intervention would never be so obvious and directly observable. Everything that happens in life can be explained by science and science is merely the medium on how God can do His thing and work his wonders on us. How God can intervene will be further explored in the later chapters regarding Divine Intervention.

For this chapter, we are only going to explain how the Divine Power could have worked and been the ones that dictate our life and the universe as a whole. We will also explain quantum entanglement more on how it is related to prayers and solemn request to the Divine one.

In conclusion to the first chapter, we can understand that God is indeed capable of to be in many places and be of different states instantaneously as explained by the superposition theory. This quantum theory is extended into quantum entanglement that shows that God's other ability in receiving information and getting his intervention to be manifested without the needs of any communication or information transfer.

Quantum superposition and quantum entanglement also confirm how small human beings are in this universe and there is a conceivable explanation on how there is indeed divine power that transcends all of us and how this transcendence allows action and reaction in our universe and the living things residing in it.

When you look at the sky at night you may gaze at the sky and see just uncountable the numbers of stars there are, most or some of the stars that we conceived are a galaxy itself that contains millions of planets and solar system.

When there are millions of solar system available in just one observable star that we look at the sky, imagine all of the other possible planets existing in each and every one of the stars that are in the sky that we can see with our own eyes.

That is again, only with the stars that we can see without using the telescope, even at the dark of the night where the starlight is imperceivable to us, there are more stars and galaxies.

When we look at the superposition concept, each of us are only the size of an atom in the whole universe, it makes sense that the law of superposition is applicable on us human in the universe scale- it's just that we didn't have the capability to realise it.

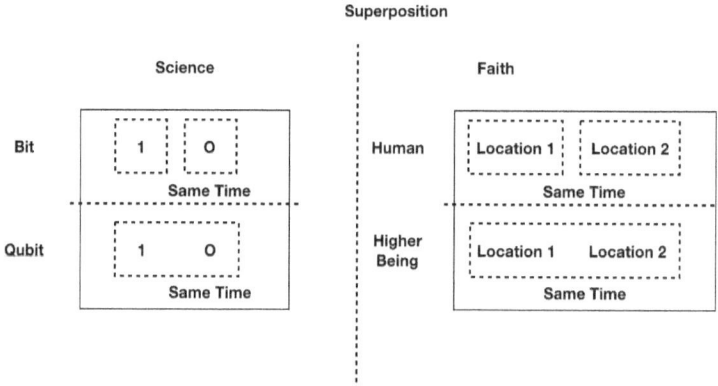

Chapter 2. Time

Everything in our life, even when speaking at a massive scale - the universe- is governed by time. When looking at the Earth itself, there are many evidences and discoveries that implied that Earth itself is quite an old planet. Some of the discoveries when being tested in the lab, the age of the discovered objects are of millions of years. Even the huge, gigantic trees that are available on Earth shows that the Earth has indeed been around for extremely long time.

Time itself is special, it is the 4th dimension of the universe: The first dimensions are dots (something like an atom), the 2nd dimension is anything that is covering the vertical (x-direction, height) and horizontal (y-direction, length) area of our visual view, the 3rd direction is when distance (z-direction) is being considered.

The 4th dimension is the time that is governing the object when moving and covering a distance. The 4th dimension is in turn being governed by another force which is the 5th dimension, which is gravity. Gravity can make time to behave differently- a planet with stronger gravity than Earth experience time slower than Earth- and that is something that human being is trying to understand and learn how to put it into use.

The most famous application would be time travelling. For example, when a person is travelling at a speed faster than light, he would be going into the past due to the fact that he will reach a place he is heading faster than the normal time would have passed. Let say person A is going to location B and it took 1 hour to go there, when person A is travelling faster than light person A will realise that person A is going back in time because he has to compensate the time where light needed to get to location B.

A simple analogy would be, when a coaster is travelling at a straight line, it should be travelling from starting point to ending point in a 1 minute, but what if it reach the endpoint in 30 seconds? Hence the roller coaster needs to have the track diverted and it needs to roll back the track until the time has come for it to go back to the endpoint and reach by the time the clock is ticking at 1 minute.

The previous story would be an example for someone that is travelling back in time, what about time travelling towards the future?

This will include time dilation and warping with a little help from gravity. When someone is travelling near to somewhere where time dilation is strong such as blackhole, he will be experiencing time slower than normal people do on Earth. When he travelled back to Earth, he would find himself to be in a world where time has passed so much- he has only aged 1 year but the Earth he's returning to can be 50 or 100 years older.

Time dilation is the product of the famous Theory of Relativity by Albert Einstein that involves the difference in time experienced by the different object when in two different situations. This situation is relative to gravitational force and the strength of the potential energy being exerted on the object.

The analogy of this would be a big sheet when spread on top of a hollow table that resembles universe, try rolling a ball across the sheet and the ball will go across the sheet without too much of a time needed - this resembles the movement of a person under normal time. Put a huge ball onto the sheet, and there will be curves around the huge ball like some kind of a 'hole', this represents gravity in the universe. Now let's put a small ball and have it rolled across the sheet again, the small ball will roll around the 'hole' before managing to go across the sheet. By the time the small ball gets across the sheet with a big ball, the small ball could have gone a few times around the sheet if the sheet is without the big ball. It is like the small ball is 'resting' while travelling along the curve and time did not pass while it is in the curve.

Meaning to say, an object that is experiencing less gravity or no gravity would have experienced time more rapidly than the object the that is experiencing strong gravity, much akin to the ball that is 'resting' in the curvy space of the sheet and does not move rapidly as when travelling on a clear sheet.

To give a relation between speed, time and gravity, a simple analogy would be again the roller coasters. A roller coaster must pick up a certain speed and by having enough speed and faster than the force of gravity, then the roller coaster would be able to defy gravity and go along the tracks.
This example also shows how gravity is involved in the whole system, much like how time can be manipulated by making use of the gravity available in the universe. For example, diverting the track of the roller coaster so that the ride will be longer by defying the gravity but the return back to the endpoint requires some help by the gravity as well.

From the analogy provided, there are two theories on how time and God is related, the time that human and the universe is experiencing is faster than what God is experiencing. In another word, in God's time may be only one day has passed but for human beings on Earth, hundreds and thousands of years have been experienced. This may explain why God is always around, has always been and will be for an infinite amount of time conceivable by a human.

However, there is another theory, the time frame that is created only works on the universe, God is not governed or affected by the power of time. For example, a vinyl or a DVD of a movie has its time frame, and every content that is in them can only be existing and working at that much of time. God, however, is greater than that and is not in the vinyl or DVD, hence is not part of the time system.

Let's get back to the roller coaster analogy, when a coaster is rolling on its track, it can go back to the same track that it has passed but it will never go beyond the starting point of the track. This is how our life is governed. We are governed by this system where we can only go so far but not back at the starting point and also we will only end up at the end point and not more.

In 2016, Laser Interferometer Gravitational-Wave Observatory (LIGO) detected gravitational wave that confirmed that time indeed can be stretched and manipulated when something that is strong such as gravitational energy is exerting its force. LIGO basically is measuring a transmitted wave and in 2016 they detected that there is a manipulation in the data in which the time for the waves to arrived is delayed which proved that they have been 'pulled' or experiencing impact and consequently made them travelling slightly further and slower to their detecting end point.

This wave is behaving exactly like the roller coaster, instead of rolling straight from the start point to the end point, it is diverting its course and took some time for it to return back to the original route and get back to the finishing point.

A gravitational wave results from the merging of two blackholes creating a supermassive blackhole with a very strong gravitational pull that disrupt the universe like a stone being thrown into the water. When there is disruption, ripples are made and propagate across the universe that eventually detected by the detector in LIGO.

This is a case to show that when something massive like a supermassive blackhole is being conceived, the universe experienced disruption and in this case the gravitational waves will be produced and could disrupt time by expanding it or allowing an object or a person to experience time differently and affected differently by time.

By this, it is safe to say that the presence of God is independent of time, or if there is any, the effect of time is highly negligible. In a way, God would have understood a person's past and how it will affect the decision made today and how the past has made a person who he is and how is he living his life.

There is a benefit in knowing a person's past due to the fact that you are more understanding of the decisions and actions taken by that person. It's the same with someone who is older and wiser, he would be more calm and composed when being presented with hardship in life and difficulties of the moment because most of the problems are somewhat familiar but only with a varying level of complexities and the people that are involved.

It is easier to explain what is happening and we are more understanding on why people are behaving in such a way and why the situation has come to this, while we are less sorry for ourselves and less angry on the state that we are in.

So it is important that God is not time-bounded and be able to see the whole story of mankind from their existence up to where they are in a moment, God would be the best judge and the best being to listen to any of our thoughts and complaints.

The best person to be able to be the judge of our tribulations and problems would be the one that knows what we have been through and the struggles that we are facing or having at hand. Also by knowing there is someone out there who knows and who has been listening and is still listening to our thoughts and musings, we can be relieved that there is someone who knows our story and all our confessions are not all but just buried deep within ourselves and are unknown out there.

Acceptance of one's problem and the characteristic is easier knowing the fact that there will be Divine Power out there who had known our story and background since the Divine One has been present since the beginning and throughout eternity. The day the Divine One will no longer be there is a dismissible idea.

Our life is so much governed by the time that sometimes how we measure our life is by how much time has passed and how many more time do we have to achieve or to be at where we wanted to be.

Now, when we get to step back and acknowledge something as great as time still can be conquered by God; or at least has little effect on God; then we can rest assured that we have someone who is very powerful existed that can make a difference in our life and at a certain degree, our afterlife.

Let's look at some of the examples to show that time difference does exist. One of the analogies that Einstein use regarding time passing is, one hour spent sitting near a beautiful girl feels like a minute, but a minute spent standing near the stove feels like an hour. This shows that we can perceive time differently even though we are at the same time system and gravitational energy. Things, of course, would be different when it is measured under places with different gravitational energy. One of the experiments ever conducted is called Scout Rocket Experiment:

A rocket is to fly at a height of 10,000km and once ready it is to fall free from that height to the Earth. During the free fall, there is a maser oscillator (something like a laser but in this, case it is used as a clock) and another maser was being used on the ground. When the maser on the ground is being compared to the maser from the free-falling rocket, it was discovered that indeed gravitational time dilation occurred as there is a difference of about

0.01%. This is to show that the maser in the rocket is experiencing a less gravitational force (free fall) and less time passing when compared to the maser on the ground which experienced a more gravitational force.

Another experiment called Lorentz transformation, involving measuring a clock when moving and when at rest. It is discovered that the time will always be slower when measured in resting state. When an object is at rest, the gravitational force that is exerting on the object is bigger. However, an object that is in motion is experiencing less gravity since it is experiencing a thrusting force in order to be moving.

When relating it to us and God, we human and the universe are governed by the force of gravity and we are always orbiting the star that is having greater pull, hence we are always in motion and our time is always shorter compared to the one that is in resting state.

The likely state for God to be at rest is supported by the described experiment above, in which time passing experience for the one that is at rest is slower. Hence, if God is also being governed and under the rule of time, God should be only experiencing a much slower time than us human and the universe itself.

When talking about time and gravity in the universe, we cannot run away from the word 'spacetime'. Spacetime, in a nutshell, is the combination of 3 dimensional parts of space, and 1 dimensional time. Space has height, width and length, whereas time stretch one way from the past to the future and passing by the present.

It is not easy to describe what spacetime is, but the widely accepted representation of spacetime is a cylinder with the conning part at the middle. The cylinder has its length, height and width, representing the 3 dimensional nature of space.

The top surface of the cylinder represents the past, while at the middle coned part is the present while the bottom of the cylinder represents the future. Time works as past, present and future, in one dimensional way.

Spacetime is constant and always will be functional in such way and such state. Spacetime simply exists because the almighty time itself is being governed by the spacetime, or in other words, time is just a piece of the spacetime cylinder as a whole.

As mentioned, at the middle of the cylinder analogy, is where present lies, is where all the timeline of living things are located, it's like a countless and limitless lines of time like sands in the ocean intercepting and aligning across the middle of the cylinder.

So, in order for someone to move in time, either towards the future or the past, they need to deal with spacetime. Using the cylinder analogy, when manipulating the timeline at the middle of the cone by bending the line upwards or downwards - to move towards future or past- the line would have to curve a little bit and this will be involving changing of how the timeline is stretch over the space of the cylinder. Hence, for a person to travel in time, he inevitably would need to beat the space/distance boundary.

Spacetime is, has been and will always be there as it will not experience expansion or reduction in space and elongation or reduction in time. The only thing that varies is the timeline but even timeline has their own constant boundary of how long and how short it is and will be.
To explain how spacetime has always been constant is that it does not change: human being perceives time by seeing how things around them change. Any recollection of observable changes of an object from their old state is what we call memories. The expected change that we foresee that has yet to happen is called the vision of the future. We experienced changes daily when the cells in

our bodies keep replacing itself and this is observable evidence that the time is passing in front of us.

Sometimes when we are watching a sports game without looking at the time, we don't realise the passing of time. In any game, it is only taking a small period of our lifetime, that obvious changes in time are not observable hence the time seems constant.

In spacetime, the only observable pattern is the timeline of our life. The other parts of the spacetime, which is space and time is always as it is- constant and does not evolve, progressed or depressed.

Another example where time does not change much is by shifting our attention to black holes. Black holes are known to be the energy collapse that causes itself to experience gravity so big that whatever that crossed path with them- including light- are being sucked into the black holes. It is also theorised that if someone goes into the blackhole, all the law of physics such as time is no longer applicable.

When in a black hole, time is one of the things that does not make sense anymore and those who are there in the black hole should not be able to experience changes and he will be at a constant state of now. The person will always be in the present and there is no past and future that can be experienced, perceivable or conceivable.

This is another theory of how God simply exists. God is at a place where no law of physics can be applied to him. His intervention and action, however, will adhere to the law of physics before it is affecting human or any living thing in the universe.

The gravity in the black holes is massive and simply inescapable when anything is travelling close enough to be sucked into it. The gravity of black holes is so strong that anyone that is travelling near it will experience time dilation; time passing excruciatingly slow. For that reason, a blackhole is actually the perfect tool for time travelling. A person who wanted to go to the future, let say 100 years later on Earth, should just visit and fly around a blackhole for a few hours or days and return back on Earth. But if he wanted to go back to 50 years earlier post-blackhole trip, he would then has to travel at a speed faster than light so that he would go back in time.

Both are seemingly impossible to be done, but if we utilise the discussed knowledge, with a little help from a blackhole and a technology that allows travelling at a speed of life, time travelling to both past and future can be done. Time is what we have constructed to gauge how long have we been around or in a bigger context, how long have the universe been existing.   We usually measure the beginning of time since the first known event by men: the big bang.

The time from when the big bang happened until today is what we called, the cosmic time.   Time is just a representation of mathematics as we measure it with numbers and integers.

However, with this representation, we can get a glimpse of how something can be infinite: Numbers, as we know it, has no end and can go on and on. Time is also infinite, as we can never truly know when is the beginning of time (although we start measuring time from the cosmic time), and we still don't know how far into the future will the time extend.

Time, for all of its greatness, is still just a law or power that governs the universe and a tool to define the perception of changes in life.   Above it, however, is a greater power that has authority and dominance over it: God is not affected by time's great influence.

Another representation of time with mathematics in which we call it, Mathematical Analysis.

Mathematical Analysis is the study of changes experienced by a subject with regards to time. For example, in the observation of an object, the object can be of pentagonal, octagonal, hexagonal and the shape can be obtained by adapting to the surrounding forces of nature. But originally, all of those objects are circular in shape. Again, let say there is a round-shaped cliff near the ocean after some time waves and erosion will give shape to the cliff and it will not be circle anymore.

Mathematical analysis can be used to calculate and see how much changes experienced over time on the circular cliff. This is a good representation on how time can be perceived as the changes experienced by an object after going through several moments and period. There is a mathematical phenomenon that when you divide a number with another number, regardless of which is bigger, the division will yield smaller but never-ending number. This is just a visually presentable occurrence that there are indeed things that are infinite on its capability and seemingly has no end.

A division by mathematical terms is a reduction of the size of the original value into equal parts of which it is separated. But this requires long and hard work just to see how much can it be divided continuously until the calculation cannot be done any longer. This is one of the examples to suggest just how futile it is in trying to truly understand the universe,   which in itself is humongous and great.

As mentioned in the previous chapter, there are a great number of stars in the sky and to calculate every planet in the universe is equivalent to counting to infinity. Each of the starlights seen by the telescope is of different lightyears away from us, some are so far away that it is billions of lightyears away from us- which is older than the cosmic time- suggesting that time is very much a mystery to be understood by our limited capacity while there is so much more unknown to it.

For example, how can a star be older than the recorded cosmic time, or in other words how can the stars exist before the beginning of time? This is a reminder of how time dilation works, wherein some of the galaxies black holes existed with gravity so massive that it is inescapable for anything that has entered its energised area.   The light coming from the stars, lucky enough to be spared by the blackhole, still will be affected and experienced time dilation due to the extremely strong gravitational pull around the black hole.   These affected light due to time dilation will cause the starts to be perceivably 'older' than the universe.

Another explanation is how the universe is expanding: the universe is thought to be around 13-14 billion years old. Let say when the measurement and observation are being done at the beginning of time, Star 'A' is maybe is around 7 billion lightyears away from Planet 'A'. But after 13 billion years have passed plus the expansion process experienced by the universe, the light needed to travel from Star 'A' to Planet 'A' will be longer. There will an extension to the calculation of the lightyear distance between them.

The effect of gravity, blackholes and time dilation, created a perception as if the time for light to reach from one galaxy to another galaxy to be longer than the total cosmic life.

There is a branch of mathematics that is called the decimals which basically is a study of numerals. In decimals, there is a positional representation of numbers and how the degree and speed of incremental vary with its position.

For example in a time board in a racing match, there are a few sets of numbers that showing the minutes, seconds and milliseconds of time. Each set of numbers are related but do not progress and change at the same speed and at the same tempo. It takes 60 rounds of changes in seconds before 1 minute passed, while it takes 60,000 rounds of changes in milliseconds before the same 1 minute or 60 seconds have passed. This is an easy representation and example how time and changes do not work at the same speed and same rate.

In this mathematical representation itself, we can observe that there is a lot that can be happening in 'minutes' time when in comparison with the 'seconds' which is how we can compare human being's time with God's time. In 7 days of God's time could be a thousand or a million years in a human's time.

So the infinity of God is basically due to the human's perception when in comparison to a human's life and existence.

When watching a formula one racing match, every second count as each split seconds determine the winning team and the losing team. The separation even between the winner and the runner-up can be just under 5 seconds. Many things can happen within a minute: when looking into behind the scene with a closer detail, much work is actually being done there and is also being carried out at a rapid and fast pace. This is also a way of saying that for a greater good to be happening and achieved, a lot of smaller works and little things need to be done by a big number of people and action takers.

In order for a goal and target to be met, a lot of effort and hard work needs to be executed.

Time is basically what we perceive based on the changes that we experienced in our life. We, humans, are also creatures that always crave for changes and we can be incapable of feeling content with just one situation and condition at all time.

Unlike God, God is quite naive in a sense that he is capable of doing the same thing over and over again, like a toddler who is watching the same episode of a cartoon repeatedly and still enjoying it as much or more than the previous watch.

For God to be able to see the stars and the creature to live from the beginning of the universe to the observable present, patience is needed since the process is going to be happening over and over again.

In a more understandable analogy, God is capable of seeing human trying to live a sinless life but somehow failed, but God is willing to forgive again and again till the end of the human's life.

2.1 Expanding universe

The universe today is not like what it was at the beginning of Cosmic time. There are cosmic inflations that make the calculation of the age of the stars and lightyears measurement slightly complicated. For example, the position of a constellation may be about 7 billion lightyears away from the Earth at the beginning of time, and due to the expansion of the universe, the position of the constellation after 13 billion years should be much further than 7 billion lightyears.

However, there is also a theory raised by famed astrophysicist Neil Degrasse Tyson, he theorised that there is a possibility that due to universe ability to expand bigger and bigger, and it is expanding so fast that there is a possibility that the universe could have split into halves due to the rapid growth.

This created a situation where there are multiple universes but it all came from one ever expanding and split universe. One can only imagine how big is the burst resulting from the splitting of the universe.

This raised another question.

Due to the expansion of the universe, what if the split universe collided with each other? A simple collision between two stars is so violent that it could create a supernova and huge explosion. The explosion wave itself could destroy anything that is on its ways.

There is a chance that the universes collision could create an even bigger bang and the explosion may also destroy a smaller universe. These are the things that will be hard to observe or it is beyond any generation's let alone a person's comprehension. This is an example where we identified the limitations of being human.

If we look at our world, the world is quite good in regulating itself, if the temperature is rising, it will bring flood and rain in order to reduce the temperature.

If planet Earth is smart enough to regulate itself, there is always possible for the universe to be able to regulate itself and gaining balance.

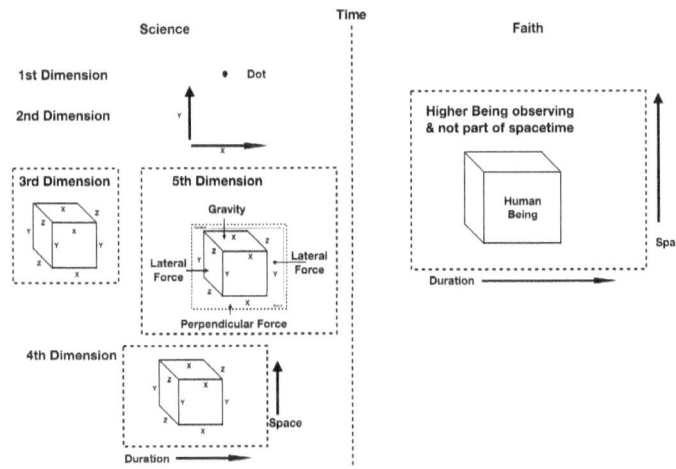

Chapter 3. Parallel Universe and Supersymmetry

There has always been a wonder and searching whether there is indeed a parallel universe to the one that we are living in. One just needs to see the size of our universe that is containing billions of galaxies and inside of each galaxy are countless more stars and systems that could have accommodate lives and species.

Based on the theory of quantum superposition, in atomic particle size, two atoms can be experiencing two states simultaneously at one time. This comes to a realisation that parallel universe indeed possible judging from the size of us when in comparison with the universe. The concept of the parallel universe, however, has always been debated since it is very hard or when looking at the theory, it is quite impossible to know that you are living two completely different lives across the universe in different galaxies or planets.

However, there is an experiment that shows that the parallel universe is possible.

The famous experiment is called the Schrodinger's Cat. In the experiment, a cat is being put into the box containing a flask of poison and a radioactive source. When an atom of the radioactive source decayed, it will be resulting in the shattered flask releasing the poison that could kill the cat.

Let's divert our attention and explore another idea regarding this. That idea is that the cat could be alive and could also be dead as long as the condition is not being observed or the box is not being opened. When the box is opened, then the condition is observed and measured to see whether the cat is dead or alive. Prophecies can be made to know what is the state of the cat when the box is not being opened yet. So the cat can be dead and alive at the same time when not observed.

But when observation is being done, the state can only be one: meaning to say as long as there is no observation, something can be of two states at the same time. We are just like the cat in the box and we can be at both state because observation is not yet being done by those who are out of the box. But when or after being observed we can only either be dead or alive. But the problem with this idea is that it is being measured in a physical way.

What Schrodinger's cat experiment has and is trying to show is that the interpretation of quantum mechanics has to be taken into account on the big scale. The previous interpretation did not take into considerations of the minute scale of the quantum particles. So when the cat is measured in a physical state, it can only be dead or alive at one time. However, when taken into account the supreme size of our universe with each living thing, the cat can be dead and alive at the same time, in accordance to the theory of quantum superposition.

Coming back to the observation idea, when you are alive you can be conscious that you are alive, but there is no way you can be conscious of yourself when you are dead. When you are dead, your state can only be observed by others. So the observer to every one of us should be of higher power and absolute supremacy, namely God.

Let's put both ideas into consideration, let say the first universe represents the alive state of the cat, whereas the alternate universe represents the dead state of the cat. But to know in what state the cat is, we have to be able to observe both universes which is impossible for us to do. We can only observe the universe that we are currently in.

By only being able to see one state of cat representing the universe that we are in, we have to concur that there should be another universe that representing another state that the cat can be in. But to be able to observe both, which is outside of human being's capacity, the only possibility is the one that has superposition ability which is God.

Scientists have been speculating that parallel universe is possible based on the reasoning of quantum physics and the existence of many dimensions or hyperspace. In order for a universe to be created, there has to be a higher state or hyperspace in which it acts as the medium. That itself shows that there is a 'universe' before the universe that we are living in is in conception.

Which brings us to a word that will be explored more now which is Supersymmetry.

Supersymmetry is a principle that is derived to enable the mysteries of the universe to be completed. In order to gain a better understanding of how supersymmetry works, it goes back to the basic classical physics and quantum physics.

Classical physics as mentioned earlier is how the normal objects being governed under a law, say gravity. To study the physical action and reaction of an object, we need to calculate the energy needed to move the object from one position into another position in the presence of gravity. Consideration of velocity, mass, on top of energy needed and the gravitational force acting on the object will be done in order to calculate and deduce how the object will be existing and working in our everyday life.

Quantum physics is when all the classical physics laws no longer applied and how we calculate an object moves from one position to another are obsolete due to the ability of an object to be at two places at one time. Supersymmetry aims to be the bridge that linking the condition where one object cannot be at two states at a time, with a situation where the same object at its minute scale can be of different state simultaneously.

So how do the seemingly impossible marriage between quantum and classical physics are being done by supersymmetry? Let's look at it at a bigger picture, there are 4 physical forces that are manifested in the universe namely gravity, electromagnetic, strong and weak nuclear forces.

Gravity is the forces that we feel every day. From the weight that we feel on our body to the fallen objects such as leaves, it is undeniable and requires no reminder of how real and powerful gravity is. Gravity is also the thing that keeps our planet Earth in place in order to get the sun in the morning every day.

The second force is the electromagnetic force which is manifested in all of the great technologies that we have and continuously improved such as electricity, internet, machines and gadgets.

The third force would be the strong and weak nuclear forces, for example, the sun that is shining on us due to the nuclear fission that consequently generating power such as heat and the rays.

All those forces are very much the great manifestation of how great our universe is and how our life is very much affected by it. Every action and work that we wanted to execute in our life needs to be in agreement with the forces, those are the forces and power that are greater than all of us. We can never defeat and go above the forces, we can only play with their greatness and try to manipulate it to our benefit and get them into making our life more convenient and safe.

But what if all of these forces are originated from one common and even greater particle? All the powerful and ever governing ability is coming from one very humble yet great particle. This particle origin that makes sense is what we call Supersymmetry Particle. All the forces known to Earth which applied to classical physics are the result of an asymmetrical release of energy from supersymmetry particle.

The analogy of the phenomena would be an explosion of a firecracker, when the sparking fire is in contact with the explosive powder, the reaction of the fire, oxygen and the flammable material release the energy. The energy released is of many forms, for example, there will be a sound of a loud explosion and explosive light but with medium intensity and luminosity.

The release of energy represents the release of the forces from the Supersymmetry particle, and the loud sound represents the strong forces such as the gravity, while the less intense light coming out of the explosion can represent a lesser in strength but equally important force such as weak nuclear force.

The strongest of it all is the gravity, as explained in Chapter 2, we discussed that gravity has the power to slow down or even manipulate time. But the manipulation is not direct manipulation, for example when there is a data being transferred, the future will not just change itself, someone in that instant has to pick up the information and take action to shape the future as desired.

This is similar to how prayer and intention are granted. Similarly to other energies such as electromagnetic and the nuclear forces, all of these are strong enough to get things moving and work done: but we will still need to harness the energy in order for it to work for us.

But before there is energy such as gravity, it all originated from the supersymmetry particle. But when we take a step back further to the start of the particle, it is nothingness. It is discovered that light can be indeed be created from a seemingly empty space such as vacuum: with condition that it is under a proper and correct atmosphere- this is a phenomenon predicted called the Casimir Effect. We will explain the nothingness and light in the later chapter.

But for now, it has to be pointed out that in the beginning, to be able to create this ultimate particle out of nothing it has to be in a conditioned manner and the theory is that this is being done by this super-intelligent being, God.

The particle then one day is being released and the 4 major energies mentioned are the ones that become the big players in our universe and our whole existence. So this goes back the previous chapter that the reason why God is not affected by the rule of physics is that in the beginning the release of the energy has not even happened yet and there is only nothingness. After the particle is created from nothingness, subsequently it will release the 4 major energies and the universe is formed by itself.

Even after the release of these energies, they do not affect God because God is of another dimension and these particles will be the communication between the first dimension up to the fifth and physics of multiple dimensions.

At the beginning of the chapter, we are talking about the parallel universe, however, the parallel universe that is being discussed and explored in this book and chapter is the parallel universe of the afterlife. Afterlife is the world or dimension that religious or people with a faith in God believe one will be existing in after death.

There are some arguments and suggestions that after death, everyone will just experience nothingness- no heaven or hell- there is no data to be conceived and processed by us anymore. However, if we were to apply science to our life and death it will give an interesting idea;

Before we were conceived and born by our mothers, we basically come from the nothingness, based on the second law of thermodynamics, an energy will never regress to its original state: meaning to say we the human are constantly changing physically and emotionally and our condition keeps changing and we will never go back to our original state. In conclusion, our state after death will be us as a soul in the afterlife. We will discuss the concept of soul even further on the science of the soul at the last chapter of this book.

Based on the earlier discussion in this chapter, we can now say that: The state of life and death can be observed; Our soul is an example of irreversible energy; The parallel life or the multiverse of the life after death would be the afterlife. Heaven or hell may not be a location per se but it is a type of dimension that is beyond our perceivable known dimension (dot, lines, space, time, gravity).

Parallel universe and multiverse are what makes our universe so mysterious and full of unanswered question. It is believed that only when full unity of the classical and quantum physics is achieved, we can find a way to break the higher dimension barriers and learn about the multiverse and how it works: However, according to Godel's Incompleteness Theorem, when one problem is solved there will always be another one to be solved. We will discuss that theorem in a different chapter.

With the example of thermodynamics and Schrodinger's Cat, it is nice to be able to say that our faith with the Higher Being is not just a neglected affair. A human being is always being observed and intervention is possible if we asked for it and we acknowledge that we are not alone and usually have sufficient capacity to solve our life's problem and completing tasks.

3.1 Interplanetary travel and lost technology

There is a possibility that there was interplanetary travel happening on Earth before, based on the historical findings around the world. In some of the books such as ancient Sanskrit and Hindu Epics, there was a flying ship coming down from the sky and coming to the Earth called Vimana. Some of the places on Earth are reported to be having materials that are too advanced to be originated from Earth or is in no way can be designed by the ancient civilisation.

Based on the Sanskrit, the flying Vimana was carrying quite a number of people and civilisation into the Earth and the visit was being received warmly by the citizens of Earth. People are usually hostile or cautious with strangers or someone that is alien to them. However, the drawings depict how the arrival of the presumable aliens was being welcome- suggesting there should be contact and communication between the welcoming Earth citizens and the arriving people in the flying ship.

Even in the 21st century, flying on a big scale with a huge amount of people is a highly specialised and tedious skill. The flying technology that was depicted still deemed improvable and not at its highest quality and capability yet. But somehow the timeline where Vimaya was recorded is so ancient: long before the first modern flying technology invention by the Wright Brothers are reported.

This suggests that there was a lost technology sharing or technology transfer. If there was indeed such encounters between the incoming alien with the native Earth residents, it raised more questions on how did they managed to come into the Earth from space and how was communication was being done. How did the interplanetary travel possible at that time?

In 2015 there is a presentation of papers regarding the ancient Indian aviation technology in the Science Congress in India showing how Vimana indeed has an advanced technology that is possible for it to fly. However, it is quite controversial due to its adaptation from the Sanskrit with no abstract evidence being presented. Some, however, backed the paper due to the writing of the scientific paper was being done by a pilot and a lecturer.

As explained in the previous chapter, the universe is so vast and all the galaxies are as small as an atom hence God can act like a Qubit across the universe. God can be present in many places at the same time. There is a chance that God can incarnate itself into a human being, and existing as a spiritual and physical being at the same time. Simultaneously, that same Physical Being can be appearing in another galaxy and planet as we speak.

There is always a saying, religion is not important. What is important is your spirituality:

One day, human beings from the Earth may achieve technological advancement that interplanetary or interstellar travel is a normal transportation activity, and you may found yourself to be in a planet where the story of their spiritual God is the same, however, the name and the structure of the religion is different.

The big idea is that religion may not be relevant in the universe scope but that does not mean it is not the thing that can keep our souls safe.

In 2017, the science world was being shaken by the news that there is a discovery of 7 planets that are similar to our solar system and the bigger news is that all the seven planets are also similar to Earth.

It is not really bonkers to suggest that there are living things just like us living in those planets. The universe is so vast and it is ridiculous to suggest only Earth populates human being. Maybe by the time we start our journey to the planets, the mankind of those planets are 100 years more advanced than us in their technology and language.

Our Earth itself has so much mystery and widely believed to be older than the calendar that we are using. We may have a lot more advanced civilisation before us that has been lost in time with no records and history books about them. This brings the notion that our Earth itself is so full of unknowns and we may have only recorded 1 percent of it.

Comparing to the stars out there, there are stars that are so much older, in fact, millions and billions of years. Some stars that we see- that collapses and turned into nebular- happened so long ago that it took billions of years to be visible from Earth.

The universe is around 13-14 billion years old, the oldest known star is Methuselah which is about slightly younger than the universe- by no means young- are such a wonder to fathom. Back on Earth, Zircon the oldest known fragment on Earth is around 4.4 billion years old. The oldest temple known on Earth, Gobekli Tepe is about

11 thousand years.   Some pyramids are also ever reported to be around 20 thousand years old

From years to years, archaeologists uncovered older and older civilisation evidence with some of them are being built by technology that is so advanced that current modern knowledge could not achieve.

The interesting part of the pyramids built in many places such as Indonesia, Mexico and Egypt are all similar to the shape of constellations in the sky, suggesting they could even identify and see constellations to build such buildings. Some even go as far as suggesting that the pyramid building nation used to be able to communicate with each other to build a similar pyramid. Coincidence is too random to be taken as the reasoning for the erection of such a massive construction.

Interestingly, most of the oldest known structures are known to serve one similar purpose- a place of worship- suggesting religion and God has always been around people's life and religion can be said as the spark to the fire of building a great civilisation.

It is not entirely impossible for interplanetary travel to be done, it is just that more advanced technology and new ideas needed to be explored on how can it be done. The basic idea is to create a machine that gets a human being to fly and go across the universe. However, with the advancement in technology, internet and cloud, there could be a possibility that human being will travel not physically, but by uploading the soft copy version of the memories and mind and get it into another planet.

It is going to be interesting as human beings will be more creative and innovative while better technology will keep coming up. Sometimes with better technologies and newer concepts, better ideas will appear as well.

One of the most interesting ideas is by storing our memories and consciousness into the light and sending it to faraway planets that are lightyears away. Light is the fastest known moving object in the universe and if we can get ourselves (body or mind) to travel at the speed of light, interplanetary travel will suddenly seem like a normal travelling adventure.

3.2 Simulation and consciousness

Interplanetary travels are actually one of the endeavours to extend life beyond Earth. Earth, at the end of the day, is still a star and will one day reaches its expiration date and human being would still need to leave the planet. But physical travel is limited due to the mass factor and also the slowness of the movement, physical travel is also limited to our capability in surviving in places that are foreign with a different environment, for example, less oxygen.

One of the ways is to encapsulate our consciousness and transport it into another world. We can even create a world or a simulation and have our consciousness live in the simulation.

In that case, we can defy our own death: technically speaking our consciousness will live beyond death and survive in the simulation world that we created and with the mind that we have uploaded into. The constraint of travelling into other planets with our limited-capability body and age-driven livelihood can be eliminated by uploading our consciousness into an artificial body. We are our own artificial intelligent beings.

Technology is getting advanced years by years, we are starting to defy the limit and extend our boundaries. There is a chance that human being will upload their consciousness into a simulated and keep it in it for years to come. Another artificial intelligent body can embark first on the interplanetary travel in search of another planet where survival by a human being is possible.

When these intelligent robots have found the new home planet that they are looking for, 3D-printed human beings can be performed and consciousness can be downloaded into the 3D printed human: With that, civilisation can start all over again.

We will discuss the idea of consciousness travelling again in chapter 10.

One of the evidence of consciousness is how our brain can produce brainwaves by reacting to a certain kind of music, how our brain can react to a certain situation and get a fast response such as 'Fight or Flight' situation, how after a stress-releasing activity our brain can produce chemicals such as endorphin.

Emotional reactions can be felt and we are conscious when we are feeling it. When we feel happy, sad, angry and scared it is something that we can realise.   But sometimes the emotion and the reaction is so fast that only when we get to settle down and look back we realised what we have done. At the end of the day, we know what we feel and we know we are feeling it.

Interestingly we can even teach robots and artificial intelligence on how to react when we are facing a certain situation. For example, Facebook's reaction is a very good mass collection of feelings and how to react when perceiving news, information or display.

In the future when the technology is advanced enough, we could have the choice to either upload our memories into the cloud or into a robot. Why is it so important for the robots to have consciousness? This is because one day, some of us will choose to have our memories to be uploaded into a robot: memories can be saved and encapsulated but consciousness is a whole different story.

Consciousness is also something that is natural to us. It was naturally ingrained into us when we are born into the world and mature into a capable-to-think being. This is one of the examples why there is a spiritual Higher Being: who have designed human and to a certain extent animal (such as gorillas, and dogs) to have consciousness. It is something that must be prepared before a fresh memory is to be put into our brain and for us to be ready to be born into the world.

Simulation is one of the possibilities of what we human beings are currently living in. Our bodies and all that we are currently experiencing could be just the virtual reality. Our souls could be our real body, while what we experiencing in earthly life is where our consciousness is.

We will discuss in detail on simulation in a later chapter of this book.

3.3 Death and purgatory

Death is inevitable. It is very real yet it is very hard to be studied.   However, much debate has been on whether there is life after death. There is some evidence in life that it is pretty suggestive that there is life after death.

One of the examples is the trees: some of them being chopped down and made into paper, some of it then being made into pages of a book. The book itself is another form of life and in that life alone it can have an impact on the life of the readers. The information that is shared, jotted down or imprinted on the paper is then to be read by the readers, the information will be received into the mind and converted into knowledge. The story that is impactful to the reader will be remembered and likely will be applied and used later in their own life.

Another example of life after death is the animals that died thousands of years ago, that is buried under the ground, slowly being buried by layers and layers of ground and the longer it is buried the deeper it goes, the stronger is the force exerted on it and the higher is the temperature to be experienced by the carcass. Long enough, the carcass will be converted into a hydrocarbon that will be drilled by the human as crude oil and used to power up human's life and industry.

Continuing with the theory of simulation and integrating it with the belief of death, death could be the time where our simulated time is over. Our earthly life is the virtual reality with a specific time. Some of us only have limited time allocation. Some of us did not do too well to last long in the simulation.

We have always heard that our life here on Earth is a test and is only temporary. What if that is indeed the case? The life on Earth is the test in a form of virtual reality where we will be judged once we have done with the simulation on all the decision and actions that we have taken across the simulation. We would be punished or rewarded for the results that we achieved at the end of the simulation deliberation.

The punishment should be the way for us to improve, hence the author believed that there is no eternal punishment or permanent residency in hell: but one day, eventually, one will go back home to heaven. The author believes that the ultimate purpose of creation is to disrupt the omnipresent evil, and to beat the evil God created the light and the world.

A human being with a higher tendency to do evil has to be guided and taught that doing good is the way to defeat the evil. But if they fail to do so in life, they have to do it the even harder way, to go through hell where the concept of time is not so applicable. We will be discussing further the omnipresence of evil and our greater tendencies to do evil than good in the next chapter. We will discuss if God is so great why did God create evil?

Having said that, one minute on Earth may equivalent to thousands of years in hell, due to the difference in time flow and gravity: so human being should not be complacent with the idea that hell is not forever. Regardless of whether our life is a simulation or not we do hope that in the afterlife when our souls are supposed to rise, we will be in a better place or phase.

That better place is what we called Heaven.

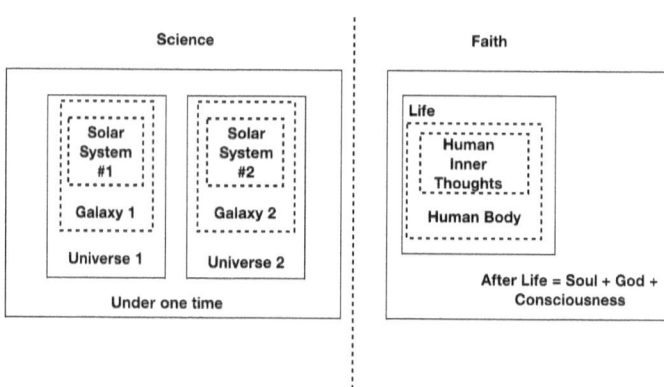

Parallel Universe and Supersymmetry

Science

Faith

Solar System #1

Solar System #2

Galaxy 1

Galaxy 2

Universe 1

Universe 2

Under one time

Life

Human Inner Thoughts

Human Body

After Life = Soul + God + Consciousness

Chapter 4. Divine Intervention

When someone is praying, the big hope is that there is someone who is listening, and when there is an intention or request for help an intervention is expected so that they can continue with their effort in carrying on with their effort in getting what they wanted or needed.

Based on the previous chapters, it is possible that intervention is being done based on the theory of superposition: God can be at two places at one time. The other theory is based on the theory of quantum entanglement, where when God is giving orders, the reaction will be us human receiving the strength or the idea on how to carry out the things that we wanted to do in our life.

There have always been debates whether our lives have long been decided and the often word that is being used is 'Destiny'. But destiny is a lazy way of saying how our life will end up and it will prevent us from striving hard to become the best that we can be. There is a better way to live our lives which is by trying our best in improving our lives and at the same time pray to the Divine Power to be blessed with courage and strong will to get it done and get things going as planned. However, our lives have the tendency to have one outcome and it is usually disastrous if we don't perform any intervention. This is called Murphy's Law.

Murphy's Law believes that anything that can happen will happen. For example, in the olden days, many prophets predicted the coming of Jesus. Historically speaking, the predicted coming of Jesus did happen: only it is not due to destiny. It happened because it can happen.

Human beings believe that they can change their future by working hard and always looking for ways to improve their life. Everyone has a plan that is already prepared and constructed for themselves, but sometimes along the way human could have gone astray and do something wrong, for example involving in criminal activities.

Divine intervention could happen by sending off signs that the person has done something wrong or the person is currently doing something that is not part of the bigger plan. Interestingly, every person has a choice: either to come back and be good again or continue with the current lifestyle and took a risk of going to a point of no return.

There is also a scientific explanation on how God can work from a place that is far from us. We are all connected and interlinked by atoms. Atomic particles are the smallest part of everything that is in the universe. All of our surroundings, seen and unseen, contains atomic particles but with space between them, that is enough for us to fit and meddle in between.

Based on the theory of quantum entanglement, even with the separation between the atomic particles with no communication between them, the entanglement is still there enough for them to react to whatever changes happening at the other side. In other words, all atoms are connected, that is how divine intervention can reach a human being.

Even though so, due to the duality of quantum, God can also intervene and get the action done through the propagation of waves. For example, wireless technology of the internet, a lot of feedback and information can be transferred across the globe by the internet wave.

Hence, the analogy for God's working medium can be as simple as the interlinking and inter-webbed internet. Through the internet, one can be connected with the news as far as the other side of the world: One could have contact with someone who is at a different country and timezone; One can get things done such as paying bills, doing business transactions, applying and completing jobs involving parties across the globe just through internet.

With internet one can have the ability to just be in the loop, to get and send communication info across, to complete jobs and making things work.

With just wave, so many things can be done. But at the end of the day, a human being still need to complete the work such as building the medium of working such as websites and mobile apps. Whereas any business transaction done online still requires delivery of stuff ordered or delegation of work and deployment of manpower for a certain job. This is not too dissimilar with how Divine intervention worked: God acts as an interfacer for any request and feedback, but the ultimate action needs to be completed and done by the human being themselves.

Everything is connected, and any puzzle of how it could work is solvable when you put quantum theory into account. For example, a seemingly empty space in the air is not totally empty, there are interlinking particles in quantum size that is invisible to the eyes. The communication between particles- even with the huge spatial distance between them- are made possible with light: Which is God's first instruction when building the universe: Let there be light. God can easily get his instructions and actions to be done in the universe with all of these interlinking quantum particles.

So the spaces between us are not really empty: all of these particles, visible and invisible, are working non stop around us. For example, God can construct the world- through the quantum particles across the universe- and slowly gather all of the information and get it transferred to the receiver particles. All of the related particles then started to combine and mix among themselves, ultimately creating the world that we are living today.

As an analogy of this, it's like how rain is formed.
Rain is made up of water (representing particles of the universe), made possible by the heat in the air (representing instruction from God) and it will make water from the sea to vaporise. The collection of the water due to heat difference will cause the cloud to be formed (akin to the construction of planet Earth).

Finally, the rain will fall back down into the ground and the sea (representation of how all of the stars will collapse one day). All of these are made possible by heat, the invisible force that created energy and temperature difference that causes rain to fall.

Coming back to the bigger picture, the universe is connected by all of the quantum particles with light as the binding force. The concept of light will be explained and discussed in further detail in chapter 7.

Let's talk about Murphy's Law again, for those of you who ever watched the movie 'Interstellar', one of the characters (a girl) was named after the law. So Murphy, the little girl one day complained to her father why is she being named for such a terrible law, then the father rephrases the law: "Anything that can happen will happen". The author prefers this phrase of the Murphy's Law, it shows that nothing in this life is actually pre-destined. It is all happening as it was supposed to be and in accordance to all of the rules that govern the universe.

Human is born because of sexual activities between the parents and how the sperm combined with the reproductive egg. Human dies because of the degeneration of the cell and body or degraded into the point of failure due to injuries or accidents.

Many of us have ever told that things happened for a reason, or it is by destiny we become who we are, or everything was part of God's plan. The author doesn't buy this idea, as explained earlier any improvisation and intervention is totally possible due to the nature of superposition and also the time relativity surrounding us and God.

Of course, in order for things to be happening actions have to be taken for it to be a reality, for example in order for a tree to grow on a land, birds or wind have to transfer the seed of the tree and have the seed landed on a fertile ground for it to grow up. The divine intervention could work in this example: If the seed is transferred by the birds or the wind too late or too early, it could have gone to an infertile ground, so intervention can be done such as tweaking the season so that the birds migration and the wind will be shifted to the time where the land is fertile.

That being said, in order for the trees to grow, further work has to be carried out. Proper sunlight and water have to be fed to the seed for it to survive and eventually grow into a big, strong tree and maybe even stayed for thousands of years into the big tree that we are seeing today.

A human can't be helped but sometimes thinking what could be happening if something else happened in their lives: what if their parents are still alive? What if they got married to that girl but not the other girl. What if they took a different career or studied in a different university. Those questions make us think of possible and alternate reality that we could have been living in.

Most of us will then resort to a conclusion that everything has been pre-destined or it has been written in the stars, especially if things and our lives have not turned out to what we have planned and desired. It is not that we are deceiving ourselves or trying to make ourselves happy, but looking at things positively is quite a powerful mechanism to let us be content in life and for us to be grateful for all the good things that we have.

There is always someone out there who has it harder or is less fortunate than us in life.

While the concept of destiny is a good way to keep us positive moving forward in life, it can keep us from fulfilling our lives to full potential. We may succumb to the thoughts that we have no time to change and achieve what we really wanted. We have taken the other road and we are in too deep to get out of the woods to switch to the other road not taken.

For a religious person, if destiny is the thing that governs the life then prayer would be obsolete and unnecessary. Why asking or praying for something that we will never have? Is it just because of our inability to see the future that we are trying to comfort ourselves, giving ourselves some kind of placebo to be hopeful that things will go our way and our intentions will be fulfilled? This is not a nice way to live and having to go through our existence.

To give an idea why this is not a good way of surrendering your life thinking it is all preconditioned, an experiment will be discussed.

Take water: Water is supposed to boil at 100 degree Celsius, no matter what day it is, which part of the world the water is, or no matter whether it is summer or winter, the water should be boiling at 100 degree Celsius.
A lazy way to put it is that it is destined for the water to boil at that temperature and one will have to be content with that nature and characteristic of the water. However, there is something else that is available in the world. There is salt. When water is mixed with an impurity that is salt, the water will be boiling at a higher temperature: more than 100 degrees Celcius.

In other words, divine intervention is like the creation of salt that can change the ability of water or its 'destiny' instead of boiling at 100 degree Celsius, it can be heated and boiling up to 110 degrees Celcius or more.

Divine intervention as explained before, still requires work to be done on our side. We will have to be the one to add salt into the water in order for the water to boil higher than its real boiling temperature.

However, there is also a misconception that everything happens because of God, and human being like to have this idea that life is always nice and pleasant. When things get awry and life gets hard or situation is not turning out as planned or prayed, then those people will suddenly lose faith in God. Some turned away saying there is no God, God is dead and some even said that for all the bad things that have happened God is an asshole. Those who also have a mindset that God is ever nice and pleasant needs to understand that God is even more complex than that. God had quite a dark side and probably quite heavy for most to handle.

There is a book written about how in the Bible, God is being portrayed as quite an angry God and even instructed some of the prophets to fight against another nation. This certainly is one of the examples that God is not all nice and pleasant: God can intervene and make your life hard and miserable, or just kill you easily and then choose to be absent for the rest of your life.

The author may sound contradictory to the next sentence but it needs to be said: bad things in life happened not because of God's plans- it is because the very nature of the universe is dark and evil.

When the light and goodness is absence from the life, evil and darkness will always triumph. That is why it is very hard to follow the good teaching of religion and the advice given by God.

However, God has always wanted goodness and holiness to triumph over evil, hence he created the light and the world so that his army will grow and beat the darkness in the end. But the hardest part is to guide human being to do so, and due to free will.

The human being has been given the greatest gift of all, free will and the ability to make a choice. Why does free will exist? So that the triumph over evil is not a forced or groomed action but out of rational and personal choices.

However, due to this provision, a human could always choose to do things that lead them away from God, for example in the famous story of Adam and Eve. Adam and Eve were being given all of the life's best stuff, freedom to live and sleep anywhere in the Garden of Eden, the ability to control the animals that are running and flying on Earth. They just need not eat that forbidden fruit that could give them knowledge: the knowledge that life is not perfect and even ugly.

The intake of the forbidden fruit was the choice to be away from God, they chose to give up the blessings of having the world as a nice and pleasant place. They chose to see the world as it really is, and the ability to see that God can indeed be hard and difficult if you chose to have him absent from your life.

Scientifically speaking, a person is given a place full of light. With refraction and deflection of light, a human being can have the vision of colour and everything that is wonderful. But out of those light, there is always darkness that we don't have to know. Light has blindspots where shadow existed, or light is just one of the electromagnetic energies that are released from the ultimate particle. The ultimate particle was born out of the darkness.

Now, back to our discussion about God and life's true colour, people have to understand that suffering is the by-product of the omnipresence of evil in the universe. Also, the ability to choose is what caused the human being to lose the Garden of Eden. Interestingly, we like to associate darkness with bad things because of the unknown and the fear of something that we cannot fathom and visualise, but at the same time, that very darkness is the true nature of the universe.   Due to the omnipresence of darkness, light is born out of it, which is created by God to defeat darkness.   In the absence of light, there will always be darkness.

Darkness is the result of the absence of light.
Hate is the absence of love.
Death is the absence of life.

When we look at a different perspective, things like darkness is always there, it just needs to be filled with something in order for it to be of something with meaning and goodness.

But why do you need to have meaning in life?
What is the purpose of meaning?
Why do people have to need a purpose in life? Even more so, why did God has to give meaning by creating the universe?
Which leads us to the creation of the world and the universe. The very first divine intervention was when God let there be light in the darkness which is the birth of the universe.

How did God create the universe from nothingness?

It is first by creating a condition good enough for the Super Particle to be made.

An example of a good condition is when 2 plates are being submerged in a container full of water. The two plates are arranged in a way they are separated and there is nothing between them (since both of them are submerged in water, water is now their surroundings). When the water is being vibrated, there is a force that causes the two plates to be seemingly attracted to each other. This force that causes attraction is the representation of the superparticle that was made out of nothing- just with a correct condition.

In a quantum point of view, under a correct condition, the quantum fluctuation was triggered and from there the creation of Super Particle, which is the very first particle was achieved; in a process called the Casimir effect.

Next, was the bursting of the Super Particle that resulted in the forming of electromagnetic waves which is light and all thing that's visible. The bursting also formed strong nuclear forces and weak nuclear forces that are responsible in the creation of planet and stars across the universe, and finally, the gravity that is responsible in having galaxies and planets structured and stayed in their orbit and places. The subsequent process would be the creation of planets which were done under very extreme condition. Planets were formed when rocks got bonded together by strong forces: such violent collision and consolidation of rocks created nuclear fusion.

Lastly, the animals that fly in the sky, swim in the sea, walk on the grounds are being made (through chemical and biological reaction) into a beautiful and perfect creation. The animals also have gone through a series of evolution depending on the environment they were living in and how well they adapted to their surroundings. At long last, the human being is being created.

There is always a debate on how human being came about, either from the evolution of the apes through the long process called Darwinism by natural selection which subsequently has the ability to survive and neatly reproduce, or the human being was created from the dust and particles of the earth. We are not going to get ourselves involved in the argument because both theories are very much applicable due to the age of Earth that is thousands of years and the long processes that we have just discussed from the creation of particle to the creation of living things.

Further example would be blackhole which is at the end of a star's life: in order for it to be formed, the star has to collapse and the pressure difference between the collapsing star and the surrounding universe caused all of the nearby particles to violently forced itself to move into the star until it burst into a blackhole. This is a representation of how our universe as a very stressful place: another reminder that the true side of our universe is the difficult part of it. Let's go back to our earlier discussion about destiny and Murphy's Law.

Life is interesting because we do not know what will happen in the future. Although we can project and predict what will happen in the future for example, such as when the temperature of the world is continuing to rise, major environmental catastrophe will happen: extreme flood and drought, but human being can intervene by changing the way we live: using renewable energies and try to use more reusable and recycled items. In the future, catastrophic events can still happen, but with our efforts and interventions, it will happen with less severity.

Murphy's Law- anything that can go wrong, will go wrong.

There is another example that we can discuss how things that can go wrong will go wrong:

Let say a person is being quarantined in a room with no windows and minimal ventilation, he is accompanied by an electric car for one hour, after one hour passed he is still safe and sound. But let say the electric car was now switched to a conventional petrol car, the car is then releasing exhaust into the room. The man will not even last one hour and will be suffocating due to the exhaust release.

Let's tweak the situation a bit: When the man is suffocating, an intervention is made and the conventional car is now switched to the electric car again. Now, when the electric car is reinstated and all of the exhaust smokes are being cleared instantly, by right the man should have increased the chance to survive the suffocation. However, the man can still pass away due to intoxication.

With this example, even though intervention has been done it is still not enough to save him from asphyxiation. Is he destined to die? Not really. The person could have done something when he was quarantined with the conventional car. Upon realising the ordeal that he was in, he could still covered his nose and mouth to reduce breathing and intake of the smoke. This is the important part: with his action, he could survive. Another intervention that can be done but with a slim chance of success is by performing medical procedure or resuscitation once he was done with experiment.   But unfortunately, he passed away.

Murphy's Law: Whatever that can go wrong, will go wrong.

As a human being with a mind capable of revisiting old memories and looking back at old times, we sometimes cannot help but remember the things that we have done wrong and experience regret. We always have this hope of being able to go back to the past and redo what we have wrongly done, or to go back in time to do the things that we should have done.   But that is not a correct mindset.

When relating to the Murphy's Law, which is anything that could go wrong will go wrong: There is nothing much that we could have done if things don't go our way or did not happen according to our plans, it is just that naturally things can go wrong and we have to live with it. We have to adapt ourselves to the unplanned happenings. There is no point really to always looking back wishing we could have done better.

We could have done better but things still can go wrong. Murphy's Law.

The synonyms to divine intervention are Destiny, Grace, Blessings, Fate. But all of these synonyms will risk human being to be complacent: they would have the wrong idea that they will always get what was meant for them to get because it is written in the stars. There is one belief that we should adopt. The belief that we will get something because we believed in it, which is called, the Law of Attraction.

When we believe in ourselves and our dreams, against all odds, we will achieve it. It is the belief that we can be the master of our own future: which is a better way of thinking because things in life are not something to be taken or given, but something to be earned.

It is analogous with the money in the bank. We may have thousands or millions of cash in the savings but does money come to us naturally or simply a given thing to us? No, we have to earn it. Even as a child when we get our pocket money, it's the money that our parents or guardian earned and they could spare some for us.

In life, to be successful, we have to see it, work for it, earn it.

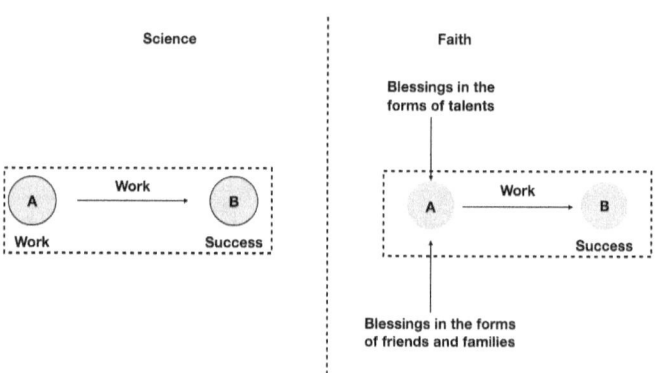

Chapter 5. Prayer

Prayer is a tool of communication between human being and God. It is a type of feedback that we give to God for deliberation. Sometimes our feedback will be compliments and gratitude which may result in more blessings. At times the feedback will be a plea for absolution and forgiveness of sins, to ask for compassion that our failings and wrongdoings. The most complicated part of prayer is the prayer of intention in which we are expecting result and intervention.

As explained in the previous chapters, the way the prayer works is not so direct and obvious. When there is a call for help, help may come or not depending on how we look at it holistically. Most of the time, we can afford to discredit divine intervention especially when it feels that we are the ones who are solving our own problems at the end of the day. But in truth, we are the ones who will need to get going and weather the storm, repair what is it that is broken in our life. We ourselves need to improve and make better of our standard of living.

Prayer is also a good meditation to the mind and soul. We have always been exposed to the knowledge of how food is good for our body: protein for our muscles, calcium for our bones, carbohydrates for our energy, fruits and vegetable for our fibre needs. But we need to acknowledge that our soul needs food too. Prayer is the food for the soul.

Of course, we can survive not consuming a certain kind of food. For example, we can survive by just being a vegetarian with no meat, or we can survive by eating zero carbohydrate. Music and travelling can be the food for the soul too, however when it comes to a really transcending experience, nothing beats the benefit of praying.

Praying can be a transcending experience due to the fact that the act of praying is a two-way action: someone giving out thoughts and musings while the receiving side of the prayer is someone who is listening, getting all of the input and gathering information. That is just the first part.

The second part would be the action taken by the One that's listening. The benefit of praying is that the person who prays can have the solace and peace in their heart knowing some of the heart's content has been released and being shared to someone else. The one that prays can experience the sense of satisfaction especially after acknowledging gratitude to the Higher Being for helping him in weathering the latest storms, by being thankful for all the blessings that have been showered.

As mentioned in the previous chapters, there are a few ways how the prayers can be made to be heard by the Higher Beings.

This goes back to the theory of quantum entanglement, in which when one person is saying a prayer, there is another connected person that is listening. This works by having particles all over us which is interlinked with light: our prayers are being captured by the light and transmitted from one particle to another until it reaches the Divine Being.

There is a movie called 'Lucy' where the main character is able to use her brain power to the fullest, and one of the things that she can do is moving things from a distance. She is not really using any superpower, what she actually did is utilising the ability to communicate with the invisible particles and atoms in the air and get the particles to move things for her. This is exactly how Divine Power can intervene in our lives.

However, God doesn't usually intervene too much because he has given us the freedom of choice. Sometimes when one prays, one can feel something in their heart resonating, deep in their core. This is because one prays, one is allowing light to be getting into them and they are releasing energy that is burdening them. At the same time, they are balancing the energy of the universe in order for action and occurrence to be happening.

The concept of prayer can be put into the analogy of cooking. When someone is cooking, following recipe and guidance- akin to a person following religious commandment or moral ethics- the preparation of the dish should be alright. A good chef or cook usually will taste the food to know the condition of the food or to know what can be improved for the dish. This tasting of the food while cooking is just like feedback to the cook: similar to with praying; When we pray, we are giving feedback to God on how our lives situation can be improved or how our hearts' desire can be fulfilled.

Things can be alright when a human being is planning a life in accordance to God's commandment, however a prayer is a medium for feedback to let God knows on things that can be done and improvement to be made.

The uniqueness of prayer is that scientifically speaking, it benefits the brain: in a way that although it is perceivably a one-way communication, it feels more like a two-way mode of communication. It is a bit like music written by composers. When they are writing and singing the song, the song touches a lot of hearts and is relatable to many people- the composer still have happy thoughts and feels that he is telling his story to people and people are loving it (with assumptions that there is no verbal or visual feedback from fans to him).

Prayer is something that we can make while we are relaxing but at the same time our brain is active, due to the language that we are using to 'tell' the receiving side. When we are praying, our brain is actively imagining things that we wanted to have and mapping out the future that we are visualising to God.

A scanning was conducted on the brain of a person who is praying as a part of an experiment, showing scan result consisting of activities in the area of the brain where language is being processed. The result is showing that when someone is praying, electric signals are being produced in the brain as if there are speech perception and production.

Prayer is also unique because it is an act of asking for something that is imaginary; in other words, something that is not real does not mean it is negligible: it still has the capability to have an impact on someone's life and on the decisions made over the course of the life.

We perceive the environment differently, we communicate with people with a different method. Human beings use verbal exchange as the means to convey messages and thoughts to the other person, while human being takes the message by listening.   How we communicate is different with how whales communicate, in water the whales communicate with a high decibel sound, they did not communicate by talking due to the different environment they are living in which is water.   A bat usually goes out at night and they can travel in darkness with no problem, how they see is not like how we see things.   Bats used waves emission for them to 'see' in order for them to go places.

These are simple representations that how we perceive things differently depending on what environment you are living in.

If we are to talk in a grander scale, the multiverse scale of God, God may be receiving our prayers differently, and how God perceive our daily life is also not through traditional 5 senses.

God 'listens' and 'sees' us in ways that are different from how a normal human being is communicating.   The point is, when we pray how we get our intentions known and how the prayers are being answered is not as direct as to how we tell our thoughts to another person and the resulting action taken.

This is also in line with our explanation in previous chapters that God can be listening to all of our intentions simultaneously and anytime due to the difference in the message receiving method God is applying. Another reaffirmation is the way God intervenes in our life: the answered prayer is not straightaway given, but in a way that requires our own action to get the intentions fulfilled. Let's enrich our understanding on this in another example, in the oil industry, the oil that we drill and produce deep underground is actually pools of hydrocarbon, that we called the oil reservoir. The oil reservoir is like a pond with rocks around it that stopping it from propagating. The hydrocarbon is initially propagating through permeable rocks called sandstones and keeps travelling until it finally stopped by another type of rock called shale which is impermeable. Over time, more and more oil will propagate into the same path and stopped by the same shale until it is big enough to be called a reservoir waiting to be drilled.

In this example, we are more interested in the permeability of sandstones. The sandstones have 'holes' big enough to let oil to run through it. But when the rock is being dug out to the surface and being inspected by our own hands, the rock is seemingly just a chunk of stones with no chance of penetration. This is a good example of how our own experience with the sandstone is different from the oil's experience with the stone.

Another analogy that we can give is radio and television. When we are listening or watching a story, with its music and its plot, we can tell which character is the hero and which one is the villain. For God, our life stories can be seen and watched like watching television. With that the outside appearance of a person can be perceived, the character of the person can be observed, and the action and behaviour of a person can be recorded.

If there is no universe filled with living things, who have desires and intentions, what is there to intervene? Well, the author may not be able to answer for the sake of God, and even if there is a really good answer given on existence and intervention, none could possibly confirm God's thinking. How God thinks may be different from how a human being is thinking.

The other question would be which prayer or what type of prayers that would be strong enough to command the divine intervention? Maybe divine intervention is granted to every human's petition; the difference between the desire and the reality is the action to be taken by a human being.

The most obvious type of Divine Intervention is what we called, a miracle. We are not going to focus so much on which case and situation that will convince God to perform Divine Intervention, but we are more interested in how the miracle is possible.

Let's go back to quantum physics.

Remember that an element which is in a quantum state can be in both wave and particle state when it moves horizontally from left to right, it can also be in a form of a wave moving towards the right. Now, miracles is when we can only see the element moving from left to right, we only see the beginning and the end, but we are incapable to see the process for it to reach the destination- we failed to see the wave part of the element which involving it moves in up and down direction towards right.

For example, a basketball is being bounced from a window located on the left end of the room and it goes out of the room via the right end of the room. Now, let's add an observer of the experiment but the observer is watching the action from the outside of the house, from his perspective, the basketball is seemingly being pushed from the left end of the room and somewhat strong enough to go out via the right end of the house. What he failed to see given his capacity is how the basket ball was actually bouncing across the room before eventually going out of the room through the window at the other end. So basically, miracles are not really miracles, it's just that we have no capacity to see what happened behind the scene.

Another similar state that resembles a lot like a miracle is what we called the Placebo Effect. Now say, a person is having a headache, a bad one, and the headache is a recurrent one. The person can't take it anymore and he went to see a doctor. Upon checking on the patient, the doctor realised that the person is indeed sick but he is out of the medicine needed to cure his patient's sickness. So the doctor comes out with other pills- just normal pills, more like sweets- and prescribed it to his patient. Upon taking up the pills, the patient suddenly experienced relief in his pain and he is no longer having headache. He suddenly is out of his misery. So, the doctor lied to his patient about the medicine, but somehow it cured his patient of his sickness.

Now, this is a miracle- a miracle of the strength of our brain. Our brain is actually a very powerful organ. It can tell us how to react- akin to fight or flight situation- and with our brain, we can suppress pain. With that ability, the author is not surprised if the brain can actually suppress the growth of a tumour by stopping the supply of important chemicals or ingredients that promote the feed to the tumours.

It is not what brain would normally do, but in fight and flight situation, our body can react and we become stronger and doing things normally we could not have done. We will be discussing the power of brain and mind in Chapter 10.

A miracle is not really a miracle, it's just the extra work that we couldn't see or explain normally.

Let's have time off and recapped some of the things that we have discussed so far so that we can apply it here. We have learnt about how superposition is an ability to be in two different states or at two locations at the same time. We also have discussed how God can be of superposition state while a human being is basically governed by classical physics and can be observed.

The next idea that we will discuss is called Wave Function Collapse.

Wave Function Collapse is when an element that is in a superposition state, is being reduced into a single state in order to be observable. In the beginning, when our souls were created we were of a superposition state in which we were not governed by normal physics, but when the souls were born into the body of human beings, we go through Wave Function Collapse and came into existence for God's observation.

So this idea fits into the discussion that we human are praying and our praying can be observed but the prayers will be received by the higher being which is in superposition. The prayers made are very much observable and did not go to waste or got lost in the sea of thoughts and other prayers.

Prayers are asking for things that sometimes we cannot explain how it can be done or how one can be deserving of it. If I am praying for my team to win a competition, and my opponent is praying the same exact thing, and God is listening to both prayers- who's prayer is to be answered?   This is now, us going back to basics: probability.

As explained earlier, things don't just happen when God allows it or grants it to happen, the baton has to be taken by us and completed by us, the human being that prays. But just how can we get it is down to how much effort we have made to get what we wanted, how well are our strategy and planning, how good and talented are we in doing what are we doing?   Taking into all these considerations give us the numbers and chances of getting the result that we wanted. Prayers are answered if we got what we asked for, and if we did not receive the ones we were praying for, then we just take it as a lesson learnt and have faith that we will do better to increase our chances to get what we have hoped for.

Now probability is mathematics that exists in our universe. By now, the reader would have reached the part that the big idea of this book that God is beyond the normal universe's law, hence probability does not also apply to God. But at the same time, God's action is still existing within the universe, hence the action has to play by the rules, for this case, the rule of probability. To simplify the discussion, the action and reaction will be translated into the existence of God.

One of the formulas that is used to calculate the likelihood of an event is called the Bayes' Theorem. The formula takes into account the first outcome (let's name it A), the second alternate outcome (we name it B). Also included is C (conditional occurrence), that A will only be possible if B is true and vice versa.

So, in 2004 Professor Unwin did a calculation of whether God existed or not inspired by the Bayes' Theorem. So the calculation involves either God really exists (A) which gives the score 1, or God simply does not exist (B) which gives the score 0. If the result of the calculation is 1 or closer to then God exists, if the result is equal to 0 or going towards 0 then God is likely not to exist. Now, C is the condition given by us. We rate C between 0.10 to 0.90, for example when God answers to prayers, that means it has a score of between 0.50 to 0.90.

After taking into all the consideration from the existence of moral and the existence of evil, the result of the probability is: there is a 67% of chance that God exists. In other words, it's likely that God exists.

Although, this calculation is not absolute; in a way, the numbering given is based on the human's judgement and the estimation will be subjective. But after taking into account all the iterations done, the amount of considerations and conditions that were taken into account, this calculation is an extensive work that deserved to be taken notice.

Not everyone knows how to pray. There are some arguments that prayer comes from the heart and the best kind of prayer is the prayer that you sincerely made in your own words. But believe it or not, praying needs guidance and teaching as well.

For example, for public speaking or even sitting for an interview: Everyone who knows how to talk supposedly can go up there on stage and give a speech in public, or when in an interview one just have to answer questions asked by the interviewer. Some people can just do it right and confidently the first time, but there are also a lot of people who can't do it right the first time, some wouldn't even dare to do the public speaking or attending interviews without guidance or proper training. Could be contributed by the fear of failure and confidence issue, or just not wanting to risk doing things badly.

The same goes with praying. It is one of the skills that requires coaching. But as a person grows in time and prays more often, the prayer will get more customised and the intention will be more personal. The quality and the level of the prayer will be getting better and mature.

In the olden days, people are inspired by the stars in the sky. Many ancient arts are showing and depicting how they are inspired by the Sun, the stars and the moon and how they go at length to create buildings and constructions that depicts the things they saw in the sky, such as the pyramids.

While it is arguable to what kind of God and deity that they are worshipping to, it is also quite fascinating and eye-opening on how belief in Deities could get people to work so hard in getting their feelings to God shown. Some of the best music ever written are some of the best forms of prayer too.

One of the most famous hymns is 'Amazing Grace', where the writer used to be a promiscuous sailor and basically living a life one will not be proud of. When he got older, wiser and also after encountering a near-death/life-changing experience, he changed his perspective of life and God. He went on to write Amazing Grace on how he was lost but am found, was blind but get to see.

There are many paintings that were created inspired by the religious event in the holy book. These are one of the examples of how life experiences can give inspirations for the creation of art- much similar to the carvings done by the ancient Earth inhabitants on the cave walls.

There are 4 kinds of learning if a robot is to achieve 'human' capability. Firstly the robot will only learn how to react: Its reaction ability is based on the instantaneous situation that it is facing but with no past knowledge supporting it.

The second stage is to be able to go through the previous history and making the decision based on the current situation and supported by past knowledge. The third stage would be when emotion comes into play, where a decision is made taking a reference to how a normal human would react. The highest level would be the stage where a robot or Artificial Intelligence (AI) is able to understand why such emotions are being felt. The consciousness of such feelings and the ability to make sense of the situation will be used to make proper decision and reaction.

Praying is one of the highest skill that a conscious being can possess. It is about recognising that there are things that are just bigger than us. It is acknowledging that when facing a problem, talking about it and seeking help is the way to move forward from being stuck. It is about recognising that we have past and history, how feelings influence our lives. We have to be able to feel grateful for what we have today as a result of yesterdays' prayers. With prayers, one can have the ability to hope for the best and sometimes dreams come true.

Being grateful and hopeful are the attributes of a conscious being. One can be grateful when the things that one have today are the things that were once being asked for but with uncertainties whether it can be achieved or fulfilled. When a person is being hopeful, that person is having limited info on what will happen in the future.   The person may have the statistics and predictions of what could happen in the situation that the person is facing, but the desired outcome is not guaranteed to happen.   So being hopeful and saying prayers are the things that can be done.

Mathematics has given us with the ability to calculate probability and historical data of how a certain outcome will happen or not.

For example, the chance of delivering a present to the receiver in 24 hours might be low due to sender has missed the delivery transport (in this case train).   By right, the sender has failed.

But sometimes, hope and prayers pushed the sender to be creative in solving the problem that he is facing while trying to achieve the desired outcome. So the sender got an idea to find the third person who is flying from the sender's location to the receiver's location.   In that case, he managed to find short-cuts to cheat time and distance. This is made possible due to the sender's problem-solving creativity, never-giving-up attitude, and the hope that the sender has in the heart.

One of the biggest difference between human beings and robots are the decision to pray.   A robot would not pray because it is a data-driven creation.   A robot will only believe that the event will either be a success or a failure if supported by numbers and figures.   One of the things about prayer is that human being refuses to give up and accept the fate that is not desirable to them.

It is through that belief also, it enables them to think of ways to change the imminent occurrence.

If a human being is able to change the outcome into the ones that they desired, they will do it in a hope that the adjustment that they make will make a difference and be successful.   If the things that they are facing seemingly impossible for them to intervene, they can resort to prayers to have the strength to face the outcome and to go through the undesirable situation.

There are a lot of arguments that robots and Artificial Intelligence can create world war and bring dangers to a human being.   This is due to the nature of the robot, which only works to achieve the desired outcome. A robot follows instructions to achieve a certain target and will stop at nothing to get the desired outcome. They will only stop when there are changes made in the instructions. The fear is that a highly advanced robot may be able to feel, work against instructions and doing things based on their emotion.

The challenge now is to let the highly advanced robot to live in harmony with the human being. A highly conscious robot should be taught how to pray because with praying one can be grateful and hopeful: By being grateful, the robot will understand that human beings have worked hard in making them a smart and fully aware being. This will prevent them from being a traitor; By being hopeful, the robots would be able to recognise that success and failures are parts of life.

Praying is not necessarily only about communication and talking to God. Prayers can also come in forms of action, in a hope that better things will happen and good things will come to us if we work on it.

When one is striving to create something good in life, efforts will be given such as going places, meeting people and exchanging ideas. All of these efforts matter in ensuring what you desired will come true. Things don't only happen when you wish for it, you have to be the executor, ensuring all the steps taken are correct and the decision made is at the best of interest.

When we pray, it is only half of the story: How will it be answered or will the request be delivered is the other half of the story. The knowledge is incomplete but we can still believe in it. We will be talking more about the unknown and incompleteness in the next chapter.

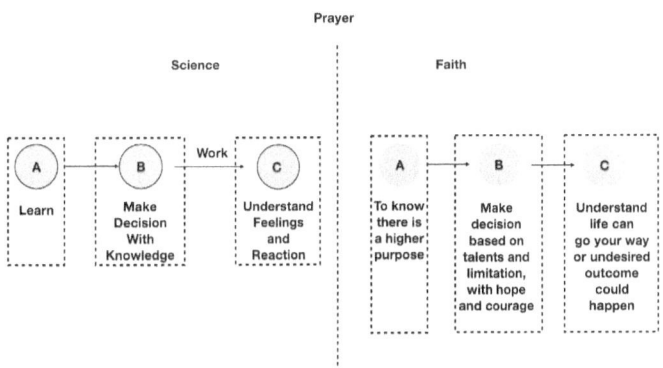

Chapter 6. Godel's Incompleteness Theorem

We have come to the middle part of the book. By now we have discussed a lot of theory and ideas. But this chapter would be the hardest to explain as we will be dealing with contradiction and recognising our human capability and knowledge as limited. Let us begin.

The existing rules and laws of Physics that govern our lives can be applicable and contradicting at times. So how is it something that is right can also be wrong at times? For example the statement: I am lying- if I am indeed telling the truth that I am lying, then I am currently bearing a false witness. But if I am indeed lying when I made that statement, then the truth that I told cannot be trusted and my confession is questionable. This is what we called, Godel's Incompleteness Theorem.

The incompleteness theorem shows that it is possible that when a law is suitable to one situation, it is still not necessarily applicable to every other situation. But when one law is not applicable to other cases, it doesn't mean that the law is not consistent and or does not belong to the established law of physics.

For example, as discussed earlier in Chapter 2, time can be dilated and manipulated by the stronger force that is gravity: the law of time is consistent within every system that has the same strength of gravity surrounding it. But after going for an interstellar journey, when one is now located at a planet where the gravity strength is much greater than the original system, based on the rule of time dilation, the person will experience time passing at a much slower rate.

Same goes to the example of one going into the heart of the blackhole, where everything that goes in has no way of going out of it. The force of gravity will be so strong there that everything will no longer be governed under laws of physic. That being said, when there is a place that the law of physics is not applicable, it does not mean that the law is wrong at the first place, it is just that a certain place does not give way for the law to work.

The Incompleteness Theorem by Kurt Godel tries to explore how the logic in our life can be consistent in explaining how the world and our lives work, but it is incomplete when looking at a bigger picture.

For us to explain the Incompleteness Theorem, we have to explain first the concept of incompleteness and consistencies.

For example, when we 'draw' a circle around the universe, everything within the circle is obeying the rules that govern it.   But if there is a circle that can be drawn around the universe, then there must be still something outside of the circle, which will be governed by another set of rules. It is wrong to say since the rules are unable to govern both inside and outside of the circle then the rules are incorrect.   This is what it means by incompleteness.

Another example, when we are giving input into a computer, we will be responded with an output. Say, if we are using an ATM machine, the input that we give is the password that we keyed in and the output will be the access into our saving account. Every time we use the ATM machine and we put in our same password, we will have access into it. Whenever we accidentally pressed the wrong password, the access will be denied and we will have to try again. This is called consistency.

But what we cannot be sure and prove, is that whenever we key in our password at the ATM machine, we will be guaranteed to get our money. We can expect access to our account but we cannot prove that there is always enough cash in the machine.   Our knowledge is incomplete.

We know consistently that we can get our money if we have access to our account, but we cannot prove that we will successfully get our money every time we got the access into our account. The machine may have coding that ensures transaction failure when there is inadequate cash available or the card that we used have reached its limit for the day. This coding knowledge is beyond us, and our knowledge of transaction with the ATM machine is incomplete.

Everybody loves sunrise and sunset. It is a beautiful time of the day when the colour of the sky is distorted and appeared to be orange in colour and sometimes purplish. The sun rises in the east and sets to the west. We can expect this occurrence every day. The Earth that we are living in is spherical, in the eastern part of the world, when the sun is getting less and less visible and the day is getting darker, the people will be experiencing dusk while the people who are in the western part of the world is seeing more of the sun as the dawn breaks in. The sun appears in the morning and disappears at night. This is happening every day- this is called consistency.

But what we cannot do is proving that the sun will come again tomorrow or the sun will be gone today. We can only expect the same occurrence to happen every day but we cannot prove that it will come. We can speculate and calculate the duration between the appearance and the disappearance of the sun but we cannot prove that our calculations and experience will guarantee the coming and going of the sun.

Let's go back to the example of the circle, if there is a biggest circle ever created enough to fit the whole universe, there is still an "outside". No inside can be created without the outside. Everything that we know has a boundary. Even the universe has a boundary. That boundary suggests that there is an outside beyond what is inside the universe. Just that it may not be something that we can fathom, there is no space, time, gravity, waves, or memories; but something that can exist in multiple forms or so foreign to us that it does not fit in the universe. Some things or rules only apply on the inside and not in the outside.

For example, when we are on Earth, whenever we drop a ball from our hand, either by releasing it from our clutching hand, or we throw it away into a direction, we can be sure that eventually, it will go down to the ground. Consistently the ball will end up moving towards the ground and settle on the surface; this is because we have gravity on Earth.

However, when one goes up to space, in the spaceship itself, when one is releasing a ball from the hand, the ball will not necessarily move towards the ground or surface of the plane. That is because the ball is no longer on Earth, there is no gravity that governs the ball as it is on Earth. The rules no longer apply in space and things are different in space when compared to a planet like Earth. In other words, whatever that we can see, hear, touch, understand are things that are in the "inside". We can use reasoning, logic, apply rules, adhere to the governance of the "inside".

But there are things that we could not understand, fathom or even imagine, to begin with, are things that are belonging to the "outside". Hence, our knowledge and wisdom are incomplete. The things that we have never seen, never heard, never tasted, or never touched but we still somehow can experience, feel and decide to believe in are what we called faith. Faith is something that we believe although we have limited knowledge of what it really is, or what is the master plan for us, and what is the grand idea behind the whole thing.

We have been discussing the concept of the finite and limited world, what about the concept of infinity? Let's try to explain it in with another concept, finite and infinite theory.

When we draw a circle, and we wanted to draw it as big as possible but as big as it gets, it can only go up to the size limit of the paper. We can also try to draw a tiny circle but as tiny as it gets, we are limited to just how tiny is the eye of the pen or pencil that we are using to draw the circle. Speaking of the circle we have to talk about the Pi system.

We know the Pi has infinite decimal numbers, or we will have 3.14159265358979... and the numbers go on. We can show the calculation of the ratio of the circumference of a circle to its diameter where it will give the mentioned numbers, no matter how big or how small is the circle, when we divide the circumference of the circle with its diameter, we will get the infinite decimal numbers of 3.141..., however, what we cannot prove is that the number indeed will reach infinity, because we have no capability to reach or write down into infinity.

This is what Godel's Incompleteness theorem is all about: No matter how consistent is the outcome, it is incomplete.

The same goes to the system beyond the universe, we all accept that anything that is beyond our universe is infinite and beyond our comprehension. Interestingly our universe for all of its glory and greatness but is still widely acceptable to be finite, but within it, there is a system that we see every day that possessed the characteristic of being infinite: Numbers and mathematics.

Hence, we have to also accept that beyond the universe there must be something that is even greater than mathematics. Which brings us to back to the analogy of the circle: What is beyond the circle?

God is that one thing that is beyond the circle, the One that created the universe. The circles that we can observe will not be possible if there is nobody to draw the circle at the first place.

We have explained how the universe is created in the quantum point of view in Chapter 4 but for this chapter, we will be explaining it in traditional science. We will be discussing both the idea of Natural Selection and Intelligent Design and how it is actually complimenting each other.

We have discussed how the One that is beyond the circle is the one that created the circle. Let's tweak the idea a little bit. Let's draw two similar and identical circle on two separate pieces of paper. The first circle will be just left unattended at one side for a few months, while we invest more time on the second circle by drawing and colouring all that we can and fit it into the second circle. We can even put some Lego, plasticine, pebbles, marbles, anything to make the second circle become beautiful.

After months have passed, we will observe both of the circles. Both of the circles won't be empty. The first circle that was left unattended will be darker or not as white as in the beginning- it has amassed dust and suspended particles in the air. Spider webs can also be formed inside the circle which is interlinking from edge to edge. Although it has been left unattended the first circle still experienced events over the few months. The second circle is more lively and colourful with all the different types of object being built onto the circle. The drawings on the circle make it looked sophisticated and good.

So let say the first circle is the world where there is no divine intervention and no Intelligent Designer, a purely natural and self-created world. The second circle would be a world where divine intervention is possible and there is intelligent designing involved in it. So it does not matter whether there is an Intelligent Design or Natural selection is what happened, our universe is somehow will get itself built into a complex system. But what is important is that there is indeed something beyond our universe.

Godel's Incompleteness theorem in a way is trying to tell us that life is full of mysteries and there will be things that can't be explained because we just have no control over everything.

For example, we have long been thinking that whatever that goes into the blackhole in outer space, will never be seen and comes out again due to the massive gravitational pull inside it.

However in 2017, astronomers have caught images via telescope showing the blackhole emitting or 'vomiting' lights out of the blackhole. This is something new and although we have consistently observed that anything that goes near to the blackhole will be sucked into it, but the new discovery shows that with the consistent knowledge that we have learned, our studies on the blackhole are still incomplete.

A visible example and experiment that we can use is our calculator or even any computers that have calculating applications. When you are dividing 0 with 0, you will always get an error or you will break the computer. No matter how fast or advanced is the computer, you will consistently get an error for dividing 0 with 0.

This is an interesting experiment that is very simple and easy to be presented for all. But it does give a question that how could calculating computers and machine, with its advanced technology are still having problems to come out with a solution of 0 dividing with 0. It is not like you are dividing 0 (a number) with something of a different 'species' like the mathematical function of 'x' or 'multiply. For that reason, you can say that the two subjects which are of not the same 'species' cannot be compared to each other. But 0 is the same exact number with another 0, which is of the same type, consistent similarity and class. The question is, why did when 0 got multiplied with 0 the machine is going to get 'error' for an answer?

Godel's Incompleteness Theorem is one of the ways to explain that there are things that we may not understand and we cannot explain but we can choose to believe. We cannot tell for sure that we will get what we want and we can know how long we will live, but we will always need to take chances hoping that we can live a fulfilled life and meet the purpose of our life.

Godel's Incompleteness Theorem gives us a reasonable faith that there is a higher being that is in control and created the universe. Understanding higher being and the universe will make us strive to keep learning and with every knowledge, we will keep feeling in awe of the things surrounding us and above us. Many times we questioned why things happened to us or why things did not happen to us. We are living in limited knowledge of what will happen in the future and what is happening around us. Things that we wanted may not come to us but instead, it goes to another person.

We sometimes wonder why are we born in a certain situation, country or ability. We just have to be strong to work our way into the desired life, although the future is unknown. We can only visualise and project what can happen in our future but we understand that things can change because life is not revolving only around us but there are many people who are living all over the world maybe even the universe.

The Author would like to reiterate it is okay to believe in something even if you don't really understand it. We can say the same for faith and beliefs, many people may argue that things that cannot be explained are things that are not worth believing. But that is a very lazy way to negate someone's faith and beliefs.

We have to recognise that we have limited knowledge of many things. We cannot even fully know how big the universe is, and whether which planets and galaxies contain another living being, we cannot experience things that are of higher dimension. These are some of the pre-requisites in showing that God is something that we cannot understand just by using our limited human knowledge and capability.

There are a few more questions that we could raise.

When living things are created, why are we created not equal, or why do we are created with limitations? Imagine all living things created with collective consciousness and intelligence. With that, everyone would have equal intelligence and can be equally conscious and will be sympathetic and have greater empathy for each other.

The speculative reason why collective consciousness and intelligence were not the way of our existence is to enable each and every living thing to have their own individuality and character. With that as well, human beings are able to make choices for themselves or for others who are of their interest or related. For our planet Earth at least we are not operating in any centralised intelligence but we all learn how to be intelligent and taught on how to make better decisions.

Imagine a world where living things can operate and live individually and independently but with collective consciousness and intelligence- the world will be peaceful and successful due to everyone is having equal intelligence and also an equal sense of importance towards themselves and the people around them.

When artificial intelligence is at their height and robots are at their peak in terms of advancement and technology, collective consciousness and intelligence might be the best way to keep robots and human beings safe.   But even we human beings cannot really predict what we can do and will do under different situations or repeated circumstances, hence it is harder to ensure robots and artificial intelligence will always work like how they were instructed.

To recap again on Murphy's law, whatever that can go wrong will go wrong, robots that have gained consciousness and ability to feel and think would be even more unpredictable and can go rogue.

At the same time, we cannot say for sure that things will go wrong for artificial intelligence. We can ensure safety by having the ability to control the robot or reset them when things go wrong, but by having the knowledge unknown to them.   This is how we can use Incompleteness Theorem to our advantage.

Things can be consistent, but incomplete. We need to use it to our advantage instead of ruing on our own limited capabilities.

Let's have a quick revisit to the multiverse idea from the previous chapter and apply this chapter's theory with it. When a person is rolling a dice, the dice are capable of only showing one side of it to the observer, however, the dice can be showing the other 5 sides of the dice in 5 other universes simultaneously. In other words, we could be only seeing the fraction of the dice, only one side of it at a time at a universe.

Although the idea is a bit hard to accept it is consistent with the Godel's Incompleteness Theorem which means no matter how many times we roll the dice, we will only see one result, it is consistent but it is incomplete, we cannot see the other sides of the dice instantaneously. It shows just how little we know about the universe and how little do we use our brain out of the full capacity and capability.

This is also a result of observer effect. The universe or multiverse can have many scenarios at one time and it can happen when there is no observer. Once there is an observer, it can only produce one result, meaning to say although the observation is not wrong it is limited and incomplete due to the limited capacity of human beings to perceive, process and experience things.

There must be a day where you wonder why the human being exists, why are the world and the universe created, what is the meaning of life. But before trying to answer the question, let's relate to our own experience. We are at our happiest if we can share our experience with someone else. If we are sad we would like to talk and tell others about how we feel and what is letting us down.

When we are creating something nice such as art, be it songs or painting, we will always have in mind to show it to our friends and our supporters. For big clubs and sportsmen, one of the biggest motivation to perform better is to make the supporters happy and to make the fans proud.

For artist, singers or performers, one of their biggest passion is to perform in front of people, for every performance they will give their best in order to get the positive reactions from the audience and those who came to see their concert.

Singers and bands will always be excited to share the songs that they have just written or just composed, sometimes because of business restriction they have to refrain themselves from sharing it, so what they do is they share the snippets of the songs.

For photographers and videographers, they are the chasers for magical moments and beautiful sceneries. Sometimes they go length just to capture what they are hoping to capture or sometimes they recorded something by accident. Once they realised that their work is worthy to be produced and released for the world to see, they will make sure that the product is really ready in order the world get to see the best version of the work.

When it comes to the relationship, the best feeling is when we can share our feelings with our significant ones, our family members or our best friends. We would like to share our joy on the happy news. When we are facing obstacles in life, it is nice to know that there are people who are willing to listen to us and share in our sadness and tears. For lovers, it is always nice to be able to share the feelings, to love and to be loved.

Our ultimate goal is to find our significant ones for us to share our feelings, love, worldview, dreams, home and future, to create a family and raise children. It is in our nature to share. The same can be said about the Creator of the world and the universe. One of the many reasons could be that the world and the universe are so beautiful that it has to be shared. The feeling of being alive, experience the joy of living and loving is so beautiful that it has to be known by many.

How great is to be able to see views, listen to music, to taste the delicious food, to touch the skin of our loved ones, to enjoy little things, to do epic stuff. To share these experiences with another person is one of the best feelings that we can have.

Up until now, we have been trying to justify our existence, but we can never prove it and our knowledge is far from complete. Just like what is spelt out by Godel's Incompleteness Theorem.

Godel's Incompleteness Theorem

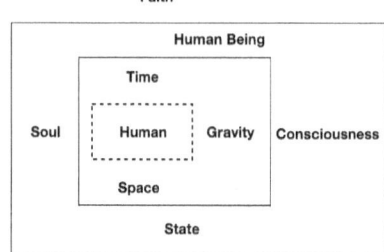

Science	Faith

Science side:
- 1.0+1.0 = 2.0
- 1+1=2
- 1.00 + 1.00 = 2.00

Faith side:
- Human Being
- Time
- Soul
- Human
- Gravity
- Consciousness
- Space
- State

Two Decimal Places only applies at the outer most layer but not at the other two deeper layers

Human body is limited to time, space and gravity, but the latter ones did not govern the soul, state, and consciousness of a person after life.

Chapter 7. The Seen and Unseen Forces

Chapter 7.1 Light

Albert Einstein suggested that there is nothing that can travel faster than the speed of light, although another theory states that when a subject is travelling faster than light it is actually travelling backwards in time.

Although Albert Einstein's Theory of Special Relativity did not forbid anything to travel faster than light but it is practically impossible due to the fact that a particle with mass needs infinite energy to be able to travel faster than light: the closer to the speed of light a particle is moving the more energy is needed, which is translated to more mass needed to provide energy. Chances are better when both light and the moving particle are in a vacuum.

In other words, travelling faster than light can only happen at times when gravity and time dilation is involved.

Light has been the ultimate tool that we use to measure everything: from travelling within time to observing events, and to manipulation of an object by the force of gravity. It is with light that everything existed and happened. In the Bible, the word that God used to start our universe is 'Let there be Light'. Our universe started when there is a big bang which is only observable and existed when there is a release of light and projection into the furthest distance which finally settling down into the planets and the stars as we know today.

The biggest theory that will be presented in this book is the concept of darkness and light. But for now, we will discuss one real and omnipresent force which is the evil and the darkness. Have you ever wonder why is it in human's nature to be evil or human being has higher tendencies to do bad things rather than being good? The nature of existence even before the beginning of the universe is darkness, nothingness and evil. Evil may also be a word that is made up just to describe an absence of good.

But what if 'good' is something created by God?
God created the universe, the light and the living things to counter the omnipresent darkness and evil. Happiness and everything that is good, are created for a better reality and existence. The knowledge that God wanted to keep from Adam and Eve is that the very nature of everything is evil and full of darkness, and God created the light and the universe so that human being can enjoy the creation that is good.

Unfortunately, human being human is attracted to their nature and they wanted to have the knowledge. But once they received it, their eyes are opened only to see that the reality is not that rosy. That should explain the saying, 'Everything that is good comes from God'.

To give an analogy, heat will travel from somewhere with higher temperature to somewhere colder which is lower in temperature. Heat is like the good thing that God has made, but it will always travel towards the colder side, the nothingness, the absence of heat. This is one of the scientific manifestations that the omnipresent condition is the opposite of what God has created- the prevalence of nothingness and wickedness. Everyone that has strong faith in religion or the atheist who is very understanding of the concept of being human, knows that human being is a very flawed being. We have more tendencies to do bad things or sinful action.

If we look at these behaviours, it seems that it is in our nature to be doing bad things, as if we are more attracted to go that way.

This is almost similar to the tendency of heat to go from a place with high temperature to a place with lower temperature. This is naturally the way heat transfer works. A civilised people would have gone to school or religious classes to know what is right and wrong: to know what is expected of the society and the religion, to live a life that contributed to the greater good, to lead a life that ensures good things to happen to you in the future through efforts and hard work.

In other words, we were conditioned in order to behave in a way that is acceptable to society and that could earn us respect and rewards.

Coming back to the heat, in order for heat to not leave a certain point, it has to be contained with insulation. But one way or another, some of the heat would found ways to leave the insulated part to go to a colder place. For us human being, we will always make mistakes and do things that are deemed wrong in the eyes of religion and God.

Why is this happening? Why, do people keep doing bad things or have more tendencies to lead a sinful life? What can we do to stop the naturally occurring rot?

The creation of goodness and light are the first Divine Interventions in the history of creation. This the reality and knowledge that God wanted to keep Adam and Eve from because the truth is that evil are all around and we human being will be easily attempted to go towards evil and sinful life.

God created goodness because wickedness and darkness should not be the only thing that exists, and there should be light at the end of the tunnel: the goodness that can be created and experienced; How the evil could be wrong and that light and holiness can prevail and triumph over the omnipresent evil.

When one is describing life, they will associate it with light. When there is light, there is life. This is basically a simple concept that people use to find other life in outer space. The picture capturing of other galaxies and nebulas are made possible thanks to the presence of light.

The interesting part is even though knowing that most of the light from the stars that reached the Earth are light years away and probably we are seeing events that happened millions of years ago- people still find it necessary and worthy to find life and planets that can inhabit life. Where there is light there is life.
A very simple concept.

To simply explain why one associates light with life is because both life and light are an entity with energy, moreover, it is visible energy. By energy, it means effort and work done. It is not like darkness where without energy and work done, darkness is still and always there.

The light itself is special, with the light we got to see colours. Colours give more ingredients to the life itself. Even though for colour blind, they still can distinguish colours such as black, white and grey- it gives variations to our life.

Human beings are made up of people of different types of skin. Some are darker, some are fairer, some having more genes from the father, some have more skin complexion like their mother.

Same goes to animals: some people chose their pets based on the animal's colour, such as red fishes and white birds. Some birds like peacock are using colour to show off their pride and something that they possessed to be boastful for.

We are also using colours to distinguish things, such as soccer teams, a company's associated colour. If you are a special artist like Prince, you will be associated with the colour purple. We also associate the colour red as something passionate, or fiery, while some other colours such as green is to be related with something peaceful and relaxing. These colour itself are having their own wavelength that can have an impact on how human beings feel and think when perceiving them.

Lights are the things that can be used to measure the passing times. Day and night, brightness and darkness are the way we use to basically realising and counting how many days have passed.

There is an effort to calculate time across the history of humanity on Earth, one of the ways is by using hourglass by letting sand flowing from one cylindrical bulb into another bulb. In other words, we can call it time, but in this case, it is a visual representation of the duration of an observable event. Some other ways are by using the sun as a time measurement.

For example Sundial, a triangle shaped blade placed on a circular flat plate. The movement of the sun will cast a shadow on the plate and represent the time of the day. Another one is Obelisk, an erection of a tower where the sun can hit on it and replacing shadow on the tower. The later the day, the brighter it gets and the shadow gets lower, and vice versa as it is getting into night time.

The next one is Merkhet, ancient way to check time passing in a longer range, it is observing the location of the stars in the sky by putting Orion, the north star as the point of reference and how far the other stars moved away from it after weeks or months.

It is interesting to see that with light, a visual tool can be used to measure time and calculate the duration of a certain period. To put it simply, with light, it will show you the way and that everything happens at the correct timing to the best of the situation.

Light is interesting due to its ability to show us objects and surroundings in multiple dimensions. For example, with a dot (1st dimension) and with the collection of dots we got to see images or pixels of pictures.

With the 2nd dimension, we got to see things in x and y direction (length and height), like the moving pictures that we see on the TV and cinemas.

In the year 2000 (The age of millennial) we are more familiar with the technology of 3D, where we got to see the object at a better representation, as we got to see the object for its height, length and width. There are more depths to the visual thing that we can see. We make use of light and shadows to better represent the objects and the corners and curves that the object possessed.

Coming into the 4th dimension, where time is in the picture, we can still use light to see how time is passing by taking the advantage of the brightness of the light and the darkness of the shadow. We can only know time is passing if we can see and the clocks and the timer that we have, regardless whether it is analogue or digital we still make use of visual representation to see the passings of time.

Going to the 5th dimension, the highest level known and perceivable by a human with the assistance of light is the presence of gravity. The easiest way to know gravity exists is by observing the blackholes, where all the light bend and the bending disrupt the space and time and the disruption spread across the universe which is being captured in 2016 in the form of a gravitational wave by LIGO.

There are more dimensions in the universe, but with the current limited knowledge and perceive-ability of human, we could only see up till the 5th dimension. Above it, it can only be felt and might be slightly out of the subject that is being discussed now which is light.

Chapter 7.2 Waves

Waves can be felt. Waves can be seen. Waves can be pulled. Waves can push. Waves can manipulate time. Such are the strengths of waves and it is not even the strongest forces known to men.

Waves that can be felt for example are electromagnetic waves, where it can heat up the meat that you are cooking in the oven. Waves can be 'seen' with the projection of sound waves like how bats do at night and how they hear the echoes back to navigate through the night. There is also hope for the blind to 'see' through the projection of echoes and acoustic way-finding.

Waves basically are sinusoidal graphs that have amplitudes and frequencies. By adjusting the frequencies and amplitudes, one can make an imaging to see something beyond their capability. For example, the MRI (Magnetic resonance imaging) or X-ray uses the wave to see what is inside our body. Even the search for oil and gas, as well as objects in the sea, can be done by sonar imaging that involved transmission and receiving of the waves.

Waves can pull you and push you, like the waves in the ocean where the tides are coming and going away from the shore. The wind in the air transfer its energy into the seawater and producing waves that you can see and at a bigger scale, surf on it.

Waves are also caused by the gravitational pull between the moon and the Earth.

The universe or on a smaller scale, the planets in the solar system are bound by the gravitational pull and orbiting around the powerful star that of the sun. Some of the stars, for example, the sun in our solar system, is so big that it curved the fabric of the space and caused the nearby planets to orbit around the sun. Some of the stars in the outer galaxy may have combined with each other and caused an explosion resulting in a blackhole that is so massive that it causes a ripple in the fabric of the space that we called a gravitational wave. In other words, we can say that there is energy across the universe and it is movable from one distant place to another.

Why is a gravitational wave is a big deal?

Human has always wanted to do time travelling and one of the keys to making time travel possible is by having the energy at a certain area to be disrupted so big that it causes the gravity to be disturbed- like throwing a big ball into a calm clear water, and the energy in the ball will cause the ball to dive into the water and causes the surface around it to bend and curve towards the ball. The ripples move shows that energy can also move across space and time in the universe. Let say we are at one spot in the water, we will always be in that spot unless there is a disruption. The only way for us to move from one area to another area of the water is through ripples, that ripples represent us travelling through space and time- only possible if there is massive energy produced to create the ripples.

So far in this chapter, we have been discussing waves. Why waves and what is it going to do with the faith in God? Waves are something that is so much involved in our life, we cannot at once run away from the wave. All the lights that we see and could not see are the electromagnetic wave.

In the 21st century, our lives are governed and revolve around radio waves that we called wireless internet. So much information can be shared across the globe and continents in real time. Such is the greatness of wave that we can know things happening in a place so far away from us and beyond our human normal capabilities.

There is a technology that 21st-century mankind has mastered that has revolutionised how information gets across. It is called the internet. The internet is such a powerful wireless technology that can allow activities to be done in a very huge distance in a very short time. An event or showcase can be broadcasted live and in real time is readily available to the masses thanks to these wireless internet technologies and capabilities.

With the internet, transactions and business can be performed at a very high speed and whatever being performed in one side of the world can be completed on the other side of the world. This is possible due to the propagation of waves and how the information and instruction are able to be transmitted over a long distance from one location to another location.

There is even the term 'Internet of Things' or IoT. This is a concept where everything from simple devices to heavy machinery is able to be connected to the internet and instructions can be transmitted and received to perform jobs. More and more technology is becoming smart and is able to be connected to the internet.

In the 21st century, anything that has internet capabilities or smartphone is the device or machine that received high demands for having high functionality.   In the age of 'Internet of Things', it is expected for all the devices and machines to be able to be connected to the internet. This means all of the machines and devices are able to communicate or to be controlled from a distance depending on how the machine was programmed to function.

The internet can give a good analogy on how prayers and intervention work. When a person is praying, it is analogous to one starts connecting to the internet. When the person is connected, the desired input will be given, such as sending emails or just opening entertainment videos. When the input has been recorded, the information will be sent across to its desired destination. Divine intervention is like the response to the input. The response could be another email to answer the earlier email, or the videos are being chosen in the library to be displayed to the requester.

Another example of how the internet can change things is the example of smartphones. Smartphones' operating system usually has to go through updates in order to receive new features and better performance, it can be done by connecting to the internet and receive updates from the developer.

This is similar to the concept of praying and faith, one can be upgraded and be blessed with good things in life, but the person still has to work for it. One should connect the phone to an internet connection, go through the updates, look through the risk and terms of the new updates, and finally request for the update. The rest is the trust that the update will go smoothly and there will be no bug in the new operating system version.

At the same time, one has to work on their storage in times where the update really messed up their phones and memory. One cannot be complacent that the updates will always be good and there will be no hiccups.

Because with any hiccups and bumps, the person might succumb to the need to look for somebody to blame for. As much as one should pray as if everything is depended on the higher being, one has to work as much as if everything is depended on man.

Chapter 7.3 Cloud

Human being across the recorded 2000 calendar years has been improving and evolving when it comes to recording information and knowledge as well in storing the information for later reference due to the degree of importance of the information.

We used to store information by just memorising as much as we can, but usually, that information is the thing that we usually apply and use in daily life, such as names, faces, places and directions. As technology progressed, human being starts to record information such as events and knowledge such as medical practices into papers and

books.   Papers and books are one of the greatest
inventions in terms of technology and knowledge
recording as well as knowledge transfer.   Even after so
many centuries, and in the 21st century, human beings
still rely on papers and books to keep information, to put
in notes, ideas and knowledge.

As the years gone by, information is needed in an instant
and covering so many distances, hence the internet is the
one that is being used to store and share information
across the globe and can be accessible at any time by any
users.

Human beings are getting smarter in getting information:
in the beginning only handwritings, drawings and letters
are possible to be recorded, humanity had push
boundaries and able to record sound and real-world
images. As the years went by, human beings are able to
capture not only images but multiple images and frames
in one timeframe which is basically what video is.

All imaginary and videos at the earliest days are captured
in films.   As the technology advanced, imaginaries and
videos can be recreated digitally. That is when binary fits
into the story.   As the computers are able to compute
more binaries and numbers, the images and videos are
able to be clearer and sharper. However, the clearer the

picture or video, the more pixels and resolutions are needed, the more numbers and binaries involved, the more memories are needed to be stored.

Storage of information progresses, from papers to filmstrips (Cassettes, VHS, USB drive). As the world needs bigger storage and meeting the demand to have memories and information readily accessible, plus with the fact of the presence of wireless internet: the newest technological achievement is the creation of cloud storage.

Cloud storage's main idea is just like the name itself, it is to store memories in 'the cloud' because you don't have to have your own hardware to keep the memories. You just need the internet to have the memories transferred from your computer or devices into the cloud storage. This is quite similar to faith.

Faith is like an internet connection, you have to be able to connect to it and recognise that there is something out there. The storage is like something that you don't possess. You are entrusting your faith by saving the memories in the cloud storage. The cloud storage is analogous to the higher being whom all your concerns go to if you let it. You can get the memories back if you make an effort to be connected to the internet and know where you keep it in and when you keep it.

Miracles can happen to you and prayers can be answered, but you got to play your parts as well.

Going towards a mini conclusion, there are many forces and powerful thing in our life that is involving science and faith are the unseen forces. But the forces that we can see in our life is also full of mysteries and there is so much more backstory of light that we would like to know more and to dive into.

For now, we will make full use of it to enrich our lives and to make sense of our surroundings.

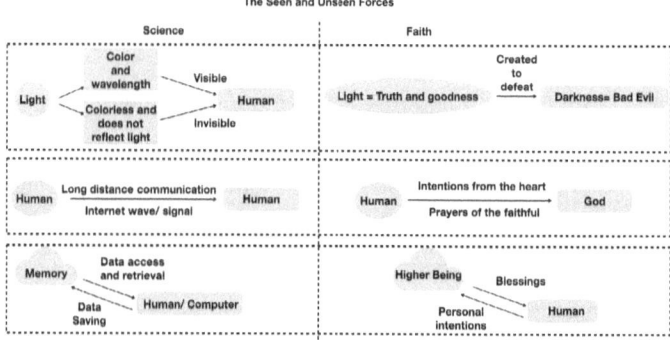

The Seen and Unseen Forces

Chapter 8. Evolution

Human beings are God's creation. Humans are made from dust. Darwinism believes in evolution. Humans are thought to be evolving from the apes and slowly becoming to what we are currently now.

Although many religions and religious intellects reject the idea of human evolution due to the belief that God should get the credit as the creator. However, both ideas are actually merge-able to be considered as one. As described in the earlier chapters, time could be working differently with God than to human being because God is experiencing different timezone under different gravity. The law of physics, in this case, do not apply to God at all due to the extreme power of gravity surrounding his presence (such as blackholes).

When God created the world, up to 7 days were used for it to be completed. However, the 7 days mentioned should not be in human's time but in God's time: 7 days of God's time could have been 70 million years in human's time due to time dilation. The Earth's age in the 21st century is around 4.5 billion years old. It is not impossible along that time evolution happens. Animals and homo sapiens (men) are all but made from dust. Both of the species fit the puzzle: Apes were first created from dust and slowly taking human being shape after millions of years. Living things adapt, and to survive the condition of the Earth living things have to evolve.

Living things have always been a survivor and sometimes adapting to changes is the way to prevent one from dying in the face of the challenging predicament. For example, human beings are facing with a huge problem of plastics that is non-decomposable and slowly stacking up and polluting the world as well as endangering animal on the planet.

However, in the year 2015, it is found that there are a new species of bacteria that could digest on the plastics: it shows that even bacteria have to adapt to new challenges to prevent themselves from dying as to get themselves more food.

Survival is human being's biggest trait, we are capable of changing our characters and behaviour just to see another day. One of the famous stories is the frontiersman named Hugh Glass, that was attacked by a bear in one of his expeditions. After the attack, his expedition mates decided to abandon and left him for dead- however, he survived being alone and not only that he managed to found his way out of the wilderness and into civilisation.

Human beings don't just have different skin colour for no reason. For example, people originated from Africa has darker skin due to the production of melanin in order to protect the skin from the scorching sun and the higher temperature of their location.

Western people have fairer skin due to their cold environment, the whiteness of their skin is due to the reduction of blood flowing to the surface of the skin in order to reduce heat loss from the body. Adaptation for survival has always been human being's greatest survival tool and it is deeply embedded in our DNA.

Human beings are also becoming more aware that life on Earth may not be lasting and Earth will meet its demise one day. In the search of the future home, space exploration is ongoing to prolong the existence of man and to have a new life elsewhere.

When talking about space exploration, we cannot help but talk about planets around us. For example, Mars has gone through many evolutions. Mars used to be a planet that is similar to that of Earth. However, due to the weak electromagnetic waves and also the destructive solar flares, the mars was reduced to a dry planet.

Planets are also can be known of its history by referring to its layers. Some of the planets used to be so cold that it is only ice. One of the traits of a habitable planet is when the planet can contain or supply water. There are oceans between the high-pressure ice sheets. Like a burger patty between layers of bread, and you can extract liquid out of the patty.

Evolution is also a trait of surviving, if we look at technologies such as phone, we can see it has evolved so many times. Phones use to be a big, heavy case containing machine that enables people to make and receive calls. Slowly phone designers are getting better at reducing useless space and with better materials, phones are designed to be mobile.

Years later human beings are getting better at making things smaller and make full use of every space there is. Processors turned into microprocessors, mobile phones with sim card changed in size into micro and later nano. With these technologies getting smaller and smaller, more technologies can fit into the phone.

The next evolution of phone is by becoming smart. They are able to connect with the internet and able to provide entertainment, perform a business transaction, and make essential jobs such as calls and messaging. Even messaging has evolved, it used to be message sending through radio waves, but when the smartphone era arrived messaging is done through the internet.

For ourselves, we can see our own evolution in our hair. Some kids used to have their hair colour to be of certain colours such as red or blonde but once they grow up their hair become black and permanently so. This is due to changes in melanin level in the hair (Eumelanin for blackness and Pheomelanin for blondness or redness). The changes are due to the gene that they have. But when they grow up and get old, their hair turns white due to the hair follicle losing the melanin. By the later stage of life, a human being would have experienced two different evolution in their hair. This evolution shows that there is a certain evolution that is involuntary and is a part of growing.

There is also voluntary evolution, take the same example hair, our hairstyle can be different. A girl can have her hair long, black and wavy. But by the time she finished schooling, she decided to cut it and have it pixie as well as dyed to red colour. After a while, maybe when she started her working and adult life, she wanted it to be straight and blond. Towards her peak adulthood, when she fully embraced herself, she decided to stay with her original and natural long, black and wavy hair. Once in a while, she will change the style just to make it interesting.

Celebrities' hairstyle usually is being documented by fans or entertainment media: From having long hair, to a dyed spiky hair, then normal smart undercut hairstyle and later continue to evolve over the years. This is an example of how evolution requires human intervention and it is a collaborative effort with the natural process in order to get the desired result. Evolution happens when one worked for it and make use of the natural process to their advantage: even when it comes to extending life and legacy in the world by defying mortality. We will discuss mortality further in the next sub-chapter.

Chapter 8.2 Mortality and simulation

Human beings don't live forever. There is only a certain length of time and years that a person can last and still be alive. Some people can live longer than the others. Some living knowing they have managed to do what they were supposed to do in their life. Some didn't live to see the fruits of their labour or get to the finishing line of their lives purpose.

The simplest analogy to explain mortality and life with God is this: basically, it involves a cook (God) that is trying to get a dish done (Our life). The final outcome of the dish is what the cook has in mind (our life's plan). The cook would be God, we are the meat and the process of preparing the meal is the journey of our life. Some journeys are more complicated that requires more intervention, and adjustment in order the life to be successful and fruitful. Some journeys are straightforward but nevertheless required proper planning for successful completion.

So the analogy for a complicated life would be like preparing a steak, a proper fire must be prepared, the position of the meat must always be changed for an evenly distributed burning some addition of oil on its surface for an optimum result and moisture. Sometimes there is a chance that the food is going to be bad or burnt if not being taken away from the fire. To relate that with our lives, sometimes there are people who die sooner than they were supposed to and people accepted that it was God's plan or how God loves them more.

A person can sometimes inevitably gone into a point of no return that the person may be lost from God's grace, in order to prevent it from worsening God could have decided to let him meet his demise to stop the rot.

As mentioned in the earlier chapters, every human being has their own timeline of how long will they live their lives and how will their lives turn out to be. Based on the spacetime model, manipulation can be done by diverting some of the timeline towards the past or the future. However, the length of the line is fixed and eventually the time will be up and individual mortality ends there.

There are many ways in extending the mortality: the most famous way is cryogenically preserved bodies. Basically, this involves hibernation but under extremely low temperature to prevent decay of the body and decomposition of mind.

The currently available way of extending life on Earth involved robotics and machine. For example, Robocop is the extension of the life of a police officer who was involved in a fatal incident but somehow managed to stay alive: by getting his body assimilated to an android body. Artificial intelligence and robotics are the future technology and it is very much possible that human being will take advantage and use the currently available technology to extending life and prolong mortality.

The signs of life after death actually are available around us. For example, the part of the trees that were being cut down, instead of being dead, were given the second chance to live by becoming the wood for the furniture or artwork such as painting frame. Some of the other woods will be given different life by becoming papers and the new purpose in life are becoming books. Some of the books went on to become important books such as the holy books, or the scientific papers that passed knowledge from one generation to another. History or story will be immortalised on the papers to be read by the readers across the years, countries and some book will be elevated into timeless status.

In 2017, one movie called 'A Ghost Story' explores about life after death, about how when a person is dead he will go to the original version of himself- which is not governed by space and time. In the movie, the ghost got to travel to the future where he got to see the house he used to stay got demolished and turned into a high rise building with its surrounding as the metropolis. The same ghost also got to go back in time and returned to the years where his house was first built which was the year where the land was first discovered.

Finally, the ghost went on to see himself on the last days of his life.

This could give a plausible explanation of how a house could be haunted, the ghost is still travelling in time around the house where the memories are still latched to the house and he has not found closure. Until he is ready, then he will no longer stick at the same space but different timeline.

It is consistent to when we are still alive, our mind is able to time travel, we get to see our memories (the happenings in the past) and look at the future (our imagination of life's plausible outcome). But our mind and consciousness are somewhat trapped in the body and brain.

When we die, our consciousness is no longer confined in a container, free to roam and go through space and time. But that also brings back the idea, what if death is a wasted opportunity if we did not manage to capture our mind and consciousness and get it to be uploaded into another world or life altogether.

There is a chance that human being will extend their lives by putting their consciousness minus the memories into the machine. The machine will be sent up to space where the machine will not rot for the next millions of years unless getting hit by outer space debris. We don't have to be there physically, just our mind and our ability to perceive things as well as making new memories and living the old memories that we stored.

In 2018, the famous real-life Iron Man, Elon Musk, and his company SpaceX sent a supercar named Tesla Roadster into space hoping that it will orbit the Mars for many years. A similar thing can be done but instead of just sending a car and a dummy, the consciousness and memories can be stored in those objects sent out into space.

There is a standalone episode of a series 'Black Mirror' called 'San Junipero' that also explores about how consciousness can be transferred into a machine where the person's mind get another simulated body and live an alternate reality: a different world with a different set of bodies. We might not even realise that we are in the simulation because the higher being of ourselves are capable to extract and store the memories and consciousness into a machine.

Another series called 'Altered Carbon' explore the ways human defying death by storing and transferring consciousness into a device called 'Stack' and locate it on the neck close to the spine of a body.

In the future when robots got smarter and the internet connections got wider coverage with stronger connections, it is possible that consciousness can be transferred from a planet to another planet. Complete with a proper human-like body already in the destination planet and human being is an interplanetary species. We don't even need to physically travel across distance, just our consciousness, memories and internet connection.

Human being is getting better in sending stuff to outer space, we will one day send just 3D printers to other planets and then we 3D print ourselves and upload our consciousness at the new planet, made possible with better internet and space travel.

Chapter 8.3 Artificial Intelligence

Robots are getting smarter and smarter by day. They are getting more and more capable to perform increasingly complex jobs. In the beginning, robots are just normal machines or devices that receive input and produce output.

As time goes by, robots are created to be more flexible in moving in x, y and z direction. They can go up, front and back as well as sideways. Back in the day, robots couldn't even walk down and up the stairs properly without toppling itself over and fell helplessly into the ground.

We used to try so hard to make a robot to be more and more human-like. For example, a human being's hand is very complex, from the way the fingers are able to move and perform jobs such as grabbing objects, typing and tapping on keyboards or touchscreen devices. In the year 2017, robots are capable enough to not only perform simple stuff such as going up and down the stairs, but they can even perform back-flip from the ground to an elevated box. In other words, they are not only smart, but they are capable to perform highly skilled activities.

Not everyone can perform backflips. There is even less than half of the world's population who are able to play, let alone win, the game 'Go'. The news that robots are increasingly able to perform complex activities is raising concern that the world one day will be ruled and dominated by these highly capable and intelligent machines.

Much has been said about artificial intelligence and the concerns of them becoming smarter and dangerous: Robots nowadays are no longer the robots of yesteryears where they are only behaving as being told or they are very clumsy at doing mundane activities such as walking and grabbing stuff.

The next thing we know, robots will take over all the complex job. In 2016, the world was shocked by the news of a robot winning the game 'Go' and beat the world champion. The win is no fluke as the robot is able to beat the expert of the game at their own game, and consequently showing the world that robots are a force to be reckoned with.

Some have already started calling for regulations to control the progress of these robots, and some are even questioning why the need to continue the initiatives and effort by the companies in getting these robots to be more and more intelligent as well as independent.

The concerns would be that the robots will one day take over the world, the first stage is that robots will get people out of job. Most of the work will be automated and some jobs will be completely independent of human's ability and intervention. Robots are predicted to be able to perform even better in completing works that human being will be obsolete in certain industries such as manufacturing.

Some are calling the threats posed by these ever growing and developing robots to be bigger than that of nuclear attacks and global war. There are so many bleak predictions of the future where robots will become so smart and powerful that one-day robots will control and rule the world. A human being can be the lower caste being and might even be the ones that are being directed of what to do and how they are going to live.

Knowing that human beings are also beings with survival instincts, human beings will rebel if robots are taking over the world and our lives. This could create a world war between human and robots.
Worst case scenario would be that there will be mass casualties around the world and human cleansing will happen. But not all hope is lost. Something can be done.

That is when the internet and Internet of Things can save us. Since in the future, every robot should be able to be connected with internet, hackers can use the internet to spread virus or bug to paralyse the robots and beat them at their own game. Robots should be connected with the internet and that will be our means to control them if they are starting to fight against humanity.

Global satellite internet is one of the great human progress in getting the world to be connected and having the internet to be ever-present. On the days where robots are fighting against human, it is best not to fight physically but fight them internally with the internet.

One of the ways is also to be able to upload our consciousness into the body of a robot through the internet and we can begin to take over them and get them to behave not in a destructive way but learn how to stop trying to take over and control the world and humanity.

That being said, it is better to prevent than cure. We could train robots to have empathy, teach them to have compassion and learn that violence is not the right way. Teach them being humane and what love is. Let the robots to be conscious that it is better to live in harmony with human beings by ensuring the world is safe and peaceful.

When we read the Bible or holy books, it is recorded that human beings used to be quite violent and can be murderous. But over the course of history, we got to see that human beings improved and slowly beginning to learn that violence is not beneficial and destructive. The intention to put violence in the book is not to show that the religion or God is violent but to show human's progress.

Human beings are flawed beings and in earlier times, murder was something that happens very often. But years after years, human beings are starting to be more civilised and humane, they even create rules and laws to protect mankind and to prevent murder and homicide. This is the evolution in the behaviour and the understanding of human regarding life.

This is the things that can be taught on robots and artificial intelligence, we should not treat them as inanimate object knowing that they have the capability to surpass human beings in strength and intelligence. Artificial intelligence should also be exposed to morality and humanity concept while trained to express compassion.

Facebook rolled out the function of giving reactions for pictures, videos or news, from there, users can react to the online sharing by giving 'like', 'love', 'laughing', 'wow', 'sad', and 'angry'. These are some of the emotions that can be taught to A.I. For something that is good or impressive, we will give approving reaction: like, and love. We expressed 'laughing' for something that we considered funny, we put 'wow' for something that is impressive or surprising to us. While for something that is bad or heartbreaking, we will put the reaction of 'sad' or 'angry'.

Other social media services also give reaction buttons for any posts and over the years, the info will be good to be shared to robots on what are some of the things that are considered as positive and which ones are negative.

What makes a human humane is that we are able to express and experience emotions. Different emotions will be induced in different situations. It comes with life experience to be able to differentiate which one is good and which one is bad, to know how much we can control ourselves in reacting to a scenario or situation.

To have consciousness is to be able to experience emotions and all the feelings that are associated with it. When the robots are able to learn to live the human moral ways, the fear of A.I. and robots to be dangerous should be able to be prevented and eliminated.

One of the things that we could not fully explain is the ability of human being to have free will and making decisions. All of us have consciousness: that ability makes us able to make decisions for ourselves and even though there might be rules or customs that could restrict us, we still have the choice and free will either to heed to the rules or break the law. Each decision may not fully define us but it does present us with risk and consequences: positive or negative consequences.

One of the examples that free-will exist is when we look at particles: Particles at quantum levels are objects that have a movement called spinning. The particles are able to spin upward and downwards but in no specific direction or angle. They are spinning at free will and not governed by any rules.

When atomic particles were first studied, it is discovered that the nucleus is at the middle of it and there are electrons orbiting around it: much like planets orbiting the sun, the sun is the nucleus and the planets are the electrons. Planets are orbiting the sun in a predictable route and spin. However, the routes that electrons take around the nucleus are ever changing and not consistent. This suggesting that electrons have some sort of free will and independence on how it is orbiting the nucleus.

This is the quantum scale representation that human beings have free will and our minds have the ability to allow us to act independently to our own consent. We are able to act randomly and becoming unpredictable. But as much as we can become random and unpredictable, we are still ultimately somewhat governed by the bigger force.

Back to the electron example, although it has the free will to move at its own route, however, it is still orbiting around the nucleus and certain rules are still commanding the behaviours of the electrons.

It is not that easy to gain full consciousness as you have to achieve certain capabilities.

For example, you have to have self-consciousness to be aware of what you want internally. You need to be aware of how you feel, either it is happiness, sadness, joy or anger. You are also aware of your personal preferences and favourite subjects, your ability to feel something internally such as intuition and guts. Sometimes those internal states are being influenced or heighten by the feelings you have.

The second level will be the ability to self-control: where you are gaining the capability to govern your own actions and reactions to any news, situations or environment. You are able to prevent yourself from acting impulsively or being reckless. You can prevent yourself from mindlessly doing something that could harm yourself or people that are related to you that can be affected by the actions.

Someone with full consciousness is able to have the motivation, the inner force that can get you hyped up on doing something, the passion that drives you to achieve something that you wanted or needed, and the emotional power that makes you not giving up on your dreams and vision at times of adversity. There are days where the things that you are trying to achieve seems impossible but deep inside you, you believed that you shall pass through the hard times and one day get what you really want.

The third level of ability is the social skill in which you need to be able to understand what you were feeling and thinking then ultimately preach it to others. The ability to influence other people into believing something and acting to what you envisaged without you forcing them to is one of the most difficult skills that even the smartest human being are having difficulty to achieve. The influence sometimes depending on your skill, your status, your subject of discussion, the timing of the discussion and the place where you and the ones you interact with are in.

Interestingly, when talking about evolution, we somehow cannot run away from mentioning Intelligent Designer or God. When we are talking about God, we will surely talk about religion and spirituality.

In the next chapter, we will discuss religion and spirituality and the impact they have on the life of the people on Earth.

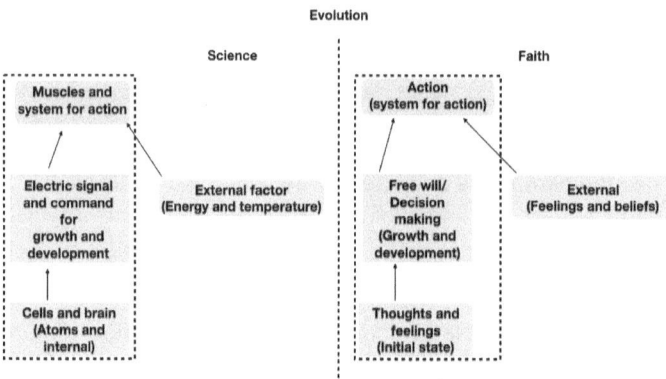

Chapter 9. Religion and bridging the World

9.1 Religion and War

There is some confusion between religion and dogma. The definition of religion is the faith that is revealed to the human beings through God or a Divine Figure. Dogma, however, is the accepted rules and decision that is made by the human with the highest religious authority on how a religion should be practised and how a certain issue should be handled.

There are also many misconceptions that most of the wars that happened in the world over the course of history are due to religion. The real reason for it is usually and mostly political. Sure, religion is the easy way out and lazy explanation to get people to support and involve in a war to defeat another person. Human has always been emotional when religion is involved due to the subject is close to their hearts.

There is a study done about how politics can be more lethal causing death via war; which is not driven by any religious motivation.

For example, Joseph Stalin is claimed to be responsible for around 42 million deaths, Adolf Hitler has been alleged to have around 20 million deaths to be associated with him, and Mao Zedong is being said to have been the cause of around 35 million of death.

The problem with religion is not that it is something that needs defending, but it has been misinterpreted as something that has to be fought for in order for all the souls in the world to be saved. The unfortunate side-effects of the concept of heaven which most religion promises are people fighting each other to get a place in heaven. Most of the religion teaches its followers that when good is being done or one goes extra length in expressing his faith, they will be rewarded in the afterlife.

One should be doing good because that is the right thing to do, not because of any reward or award for the action taken and whatever life practices. The practice of doing good things or following religion's commandment and expecting a reward (especially afterlife reward) is in a way, bribery and it makes the one who does so a hypocrite.

Based on the previous chapter it is stated that when violence is portrayed in the holy books, it is not to show that God is a violent God or favours violence but it is to portray that human beings do get better in behaviour and appreciating life.

One of the ideas that the author almost always encountered when discussing religion with friends is that the idea of religion linked with the afterlife rewards. Most followers of major religion on Earth believe or at least being taught in the concept of heaven and the final resting place where God resides and where ultimate happiness is.

The very core of religion usually is about doing good, thinking pure things and saying the righteous words. But why does one has to be taught that doing good begets good? The author agrees that doing good should be an act of itself, it should not be driven by any expectation or assumption that what goes around comes around. One of the interesting quote that the author used to hear in one of the series 'Six Feet Under' that goes something like this: "Why does one have to do good and expect to go to heaven? It's like we are bribing God to be rewarded.".

This is even more apparent for the example of a family.

We tend to agree with our friends or even strangers when they say their own parents are the best parents in the world. We knew and acknowledged that the person is confessing something that is true and real to him, we believed that their parents are really the best, but that applied in their life because they are who they are today because of what their parents have done to them.

We don't go and fight with the person who is saying that their parents are the best in the world. Although in our personal opinion, our own parents are the best in the world (even if we are to compare with other parents), but we acknowledge and accept that each our parents are the best in our eyes.

There is no need to fight for being the best.

Now take this example and apply it to religion. When a person is saying that their God is the greatest or the religion is the only way to salvation, unfortunately, there is always a reaction or dissatisfaction with the person's opinion. There is almost always a need for a person to come out and at least defend their religion and their idea of God and what that has been taught and experienced by them. Instead of sharing the stories about their religion, God and the experience that comes with it, it became an argument and debate as if the whole life and world are depending on winning the argument.

Why does it happening in religion but not in other discussions? Is it because of the idea of Heaven? Is it because the fear of if we failed in defending our religion and the need to have evidence that we fight for our religion that this is happening?

If that is the case, maybe we have to reexamine and realign with the teaching of religion that is usually about doing good and loving thy neighbours.
Fighting for religion and loving your neighbours seems contradictory although the ultimate idea is to enter heaven one day. While the idea of an afterlife has created the problem as described above, it might no longer be a problem once we have found a solution to delay, if not stopping, death.

In the earlier days, normal wounds are enough to get it infected and people get killed. But with advancement in medicinal technology, such wounds can be combated with antibiotics to cure the wounds and extend life and evade death.

Diseases such as cancer would have been a death sentence for many people, but medical technology has progressed in such a way that more and more answers and cures to eliminate cancers have been found and there are and will be more cancer survivors. Human is progressing in extending life and evading death.

More and more ideas on extending life are available: one of it is cryogenic machines that freeze and preserve your body until there is a permanent solution. Another repeating example is by uploading and downloading memories and consciousness into a virtual world or robot where you can live there forever.

Having said that, human beings are still not guaranteed immortality as we are not immune to accidents and unwanted incidents such as corrupted memories. The second law of thermodynamics fixed limits to technological innovation and human advancement in which the entropy of an isolated system is always decreasing.

However, the question is not, can we be immortal or will we ever achieve it? The question is once we have found immortality, will the concept of afterlife become obsolete? Will there be reasons to defend religion or fight for your beliefs in order to be rewarded in the afterlife? Maybe religion would be useful in guiding human being's behaviour and way of life. Religion will be good guidance on how a person, a community, a country should govern the world as well as the living and the non-living things.

9.2.1 Transcending religion: Mary

One of the most famous figures ever known in the world that is universally accepted although not a divine-being but still highly influential is Mary, the mother of Jesus.

There is a Moslem town that once was ruled by a new ruler in Portugal, and the chief of the new ruler fell in love with the local Moslem girl. The ruler then named the town after her name, Fatima, which is the most revered name by the Moslem after Maryam.

Speaking of Maryam, or Mary, the Blessed Virgin appeared in that same town in the 20th Century to deliver religious messages. That was one of the most significant apparitions and it is quite common in Europe that Moslem go and do a homage to Our Lady of Fatima. It is significant because it is considered to be the bridge that will unite Christian and Moslem one day.

If there is one thing special about Mary is that there is no one that ever denies the importance of Mary and or disregards her in the history. When it comes to Buddhism, there is a similar revered figure named Lady Guanyin, that is considered important, God's dear servant and someone that can intercede prayers to God. In the

20th century, Christianity is forbidden in Korea, but in the 21st Century, it is more relaxed and Mother Mary figure has been a unification figure on the religion between Christianity and Buddhism.

Back to the time when it is forbidden to embrace Christianity in Japan, some people resorted in a similar Buddhism figure which is Guan Yin and a child with the original intention to venerate Mother Mary and the child Jesus. There are many similarities between Christianity and Buddhism as a result of the reverence to Mother Mary. In Buddhism, chanting of songs and words are a form of prayer that is both meditating and spiritual experience; this is similar to Catholic practices of reciting the rosary which is introduced by Mary to the children of the Lourdes. Both are practices that get yourself remembering God but at the same time disciplining yourself to be committed to the prayers.

Mother Mary also teaches human to always put God first and that self-interest is not the real way to salvation and eternal happiness. This is similar to Buddhism teaching of letting go of the self and worldly needs in order to achieve peace and ultimate joy.

Mother Mary and Guan Yin also share the characteristic of being motherly, soft and tender, yet a strong figure in getting God's plan to be done for a human being to emulate. Both are the trusted figures that believers can cry to, for she is believed to have the connection and is close to God. She can tell God directly to answer their prayers and come to their rescue at difficult times.

In Pakistan, in a town of Velankanni, which is quite divided due to differences in religious view, but they come together in a festival to honour Mother Mary, or what they call Our Lady of Velankanni.

It is interesting to see there is a figure that is universally accepted and revered like Mary: That may come to the universal acceptance that a mother is someone that needs to be respected. A mother is someone who has brought life into the world and dedicated her life to take care of the children. What makes it even more interesting is that she is a religious figure whom at the previous subject of discussion, there is a tendency of people to be more defensive when it comes to religious belief and way of teaching. Mary seems to be the exception.

In a world where divisiveness still exist and people still argue on the superiority or the validity of each religion, there comes a hope that one day Mother Mary may unite us all and fill the gap of disagreement. When that day comes, Mother Mary will bridge everyone regardless of the differences in religion, and the idea of the afterlife and how to attain it.

Chapter 9.2.2 Transcending Religion: Music

We have discussed how a religious figure could unite us all. However, there is also one existing thing that is currently quite effective in uniting the world regardless of race, creed and nationality: Music.

Music has always been accepted as the tools that can transcend culture, time, language, nationality and even religion. Music doesn't have to be understood, it just has to be listened and revealed to a person for him or her to enjoy it.

There are many types of music and in each of them, are divided into many branches of categories. This is due to not everyone has the same taste of music, yet they may like 2 extreme music- very loud and very soft music but not 'in-the-middle' music such as pop music.

We human beings are also being designed in a way that we are emotional, many of life's decision are made with emotion playing big roles in it.

For example, we tend to like music that is also in sync with our emotional states. For example, when we are angry, we tend to listen to agreeable music such as rock and loud sound in order to vent out the feelings and to prevent the digestion of feeling to be buried deeply in our own heart.

Music has also been a very good way to channel our sadness and when we are feeling down. When we are sad, we tend to listen to melancholy and slow songs in order to get the sadness to be swept away by the mellowness of the song. It is also more of an agreeable sound in order for us to get over with the sadness in our heart.

Loud music is not only reserved for moments of anger. Loud music can be used to hype up a certain mood, for example before a sports competition is being played loud music are usually useful in getting the athletes' adrenaline to go up and they will be very excited to face their opponents and to win the game. Loud music is usually being used by a person who is about to campaign or to give a speech as a means to show that the spirit is high and he is about to spread it to the audience on the floor or those who are watching around the world.

Music is particularly effective in storytelling. Movies and theatres are visual storytelling, however, music is being used to invoke a certain feeling for a certain situation and scenes. If a certain scene is meant to be scary, the music that is accompanying the scene will be creepy and chilling in order to get the audience to be wary of what is going to happen in the movies. When the scene is meant to be sad or heartwarming, a heart-wrenching and soothing song are to be used to assist in achieving the desired feelings.

Some music is being composed with lyrics for listeners to enjoy. Some music is meant to have lyrics that are strong and relatable in order for listeners to further appreciate the music. The more relatable the lyrics, or if the lyrics bring a new perspective on certain issues to the listeners, good reviews will be given to that music.

Music is unique because it is the medium where we use our sense of hearing to transmit information to the brain. However, the music also involves vibration and frequency in which it affects our heartbeat when we are listening to it. When we are listening to a particular song, it can also invoke some memories of when we first hear the song or the memories that are etched in our mind that is related to the song.

Human beings are usually comfortable with familiarity, we tend to be comfortable to do things that we can do on repeat and requires less thinking but more and be in the moment and living that point of time. Studies have shown that when we are repeating something, we can expect and already know the outcome of the action or practice, and that comforting feeling can get us to concentrate with praying and focus our thoughts to God when we are praying/chanting.

For Catholic Christian, there is 'Chaplet of the Divine Mercy' where the wording "For the sake of His Sorrowful Passion, Have mercy on us and on the whole world". For Buddhist, there is chanting of "Om Mane Padme Hum', while the Hindhuism has to chant 'Mahamrityunjay Mantra" with the wording "Urvarukamiva Bandhanan Mrityor Mukshiya Mamritat" repeated for 108 times.

Although the author is a Catholic Christian, the author enjoys other religion's chanting as a way to recognise that music and song is one of the things that are created that can be enjoyed and appreciated by all. Author has also met with numerous Muslims that enjoy Christian songs and hymns because the melodies speak to them.

Sometimes songs are not just all about lyrics, it is also about the melody and how it brings joy and calmness when listening to the song. You can enjoy the song by remembering the melody or to a greater extend memorise the lyrics and know how to play it with musical instruments.

Human use music in almost all occasion, in parties, during funerals, during weddings, during religious or normal processions and during normal praying times. St. Augustine used to even said that when a person is singing during church mass, that person is praying two times- suggesting how music and singing are regarded as the highly respected actions towards God.

9.2.3 Transcending Religion: Universal communication

Tears have always been associated with the strong feelings that we have, be it happy experiences or sad occurrences. However, it has been a mystery on why do we cry. The scientist has been trying to test the health benefit of crying, however, there is no direct benefit that can be discovered. Though people do admit crying (especially due to sadness) can make them feel better as if the weight on their shoulder has been lifted up. Babies are one of the easiest examples where their only means of expressing feelings and thoughts are by crying. So crying is a form of communication.

There are quite a few things that transcend language, or in other words, universally known. Most of them are observable actions. For example one can appreciate good music no matter what language it is in, a person can appreciate a nice gesture like smiling. There is communication that is just natural to us and does not need teaching.

A person can connect to animals like horses by touching them and not making them feel afraid. A person can make everyone goes angry by doing the despicable thing like hurting someone else or abusing animals. A person can make everyone feels happy when witnessing senior citizens dancing or just having fun with their partners.

Everyone also knows that when there is a good thing that has been done, such as a person winning a competition and make the country proud, everyone will show approval by clapping their hands and give support in a form of cheering noise. Toddlers who know how to convey their feelings more than just crying are able to show that they like something by hugging or getting closer to something that they like. Most of it is some form of communication that is learned by itself and does not need teaching.

It shows that there are just things that transcend languages and cultures that we have to recognise. There are things that are just greater than communication skills that we acquired and being taught in our lifetime.

Human beings are a very visual person, almost from an early age or recorded history about human, it shows that we believe more on the things that we can see.

The earliest information of earthly religion recorded is when the ancient people are worshipping the Sun. This might be due to the reason at that time, Sun is perceived as the strongest thing that was existing that you can see and even feel of its presence and the impact of its presence. Over the years, we have seen how human beings are responsive to something that they can see and for them to pass their belief in their circles and spread it to the whole communities.

The Hindus give cow their highest respect which can be linked to how the cow is the provider of milk and food. The cow is a life provider and the source of human necessities: which is the very characteristic of God.

The Hindu God Ganesha is also known for having an elephant head and it is a representation of superiority. The elephant is one of the smartest animals, and one of the evidence is that the elephant possessed the ability to be conscious and is able to identify and recognise themselves when looking at the mirror. The elephant has a large brain and has more neurons than the human being: showing the superiority that can be associated with God.

For Christians, the death of Jesus on the cross is a visually assisted experience in explaining how God can be and wants to be involved in human beings' life as well as to prepare and reveal to them the concept of the afterlife. People will appreciate the effort more by being able to see the sacrifice made, and the action was taken by God himself and the inferior being will feel the need to be able to be grateful and willing to live a gracious life.

For Islam, the earliest commandment being given is to respect and collect knowledge of the world and the universe. By that time, books and means of collecting and documenting knowledge and evidences are already being practised. The belief is that you have to read and collect as much knowledge as possible to be able to understand how the world works and the concept of the universe. Once the stages of understanding go higher, people will be able to understand and explain spiritual things or the things that they cannot see.

It is quite universal that human being has the tendency to see in order to believe. Religion has been effectively being preached and understood by the means of visual explanation and experience.

9.3 Religion, Spirituality and Medicine

There has always been a debate about whether you need religion or not, and between the theist, the debate will be which religion is the truth and which one is the right one. The same can be said about medicine, there is always a debate about whether you need a medication for every sickness that you have. If you do need medication then the question will be which one of the medicines is the right one for you.

In truth each person is unique and coming from many different backgrounds with varied past, hence each person is lacking on something and usually not exactly identical with other people. They may have some similarities between the types of problem that they are facing and demons that they have in them, and not all or just any religion is right for the person.

Basically what is really important is the relationship between you and God, it's a very personal and complicated affair. That is due to the interaction between one person and God is not as direct as interaction with another human being or object.

It is like dangerous gases, at a certain level it can be odourless but fatal when in contact with. We can just dismiss the presence of the dangerous gas, for example, Hydrogen Sulphide (H2S), but it doesn't mean that the gas is not there and its presence can do things on you and change the condition in you. Changes usually happened very discreetly and one will only notice on the changes after the effect is quite obvious at no point of no return, or at least need so much effort to go back to status quo.

Spirituality is the acknowledgement of the authority and the omnipresence of the Divine Power without necessarily subscribe to any religious belief.

A person with strong spirituality always believe that there is hope in the face of any adversity, always have a sense of gratitude with each blessing. They believe in some kind of energy is always empowering them and how to respond to the energy is by harnessing it and make it into something positive and good in their life.

Religion is almost the same but since the mantle of leadership was taken by the human being, and to err is human, hiccups meant to happen. A human being is highly flawed living beings and sometimes it is unavoidable that mistakes and weakness will be translated into religious preaching and practising.

Religion can be said as good as trash and useless if one doesn't need it. It is like wanting a medicine when you are not sick. It is like taking a drug when you are not sick and you can end up becoming an addict.

Religion is but a medicinal drug. But it must be remembered that not everyone is healthy, some are really sick and are having a difficult life that sometimes faith is all they have: their medicine. But those who are not sick also must know and remember that one day they will be old and less than healthy: those are the days when you really need medicine.

So one must at least know the benefit of having faith or at least recognise that there is God and there are a whole knowledge and history of how God is present in life and there is an afterlife to consider.

There is a kind of people that like to shove what they believe into a non-believer's throat- they are no much different to that of drug dealers. You accepting religion just because you are told to do so is just like you taking drugs when you are not sick: there will be undesirable effects.

Not all drugs are for all sickness. So you should not be telling people on which faith and belief are right or wrong for you. Whatever that makes you a better person and let you treat everyone regardless of the differences rightly, then you are on the right path. At the end of the day we should just do good, analogically speaking we should live a healthy lifestyle/diet.

The author used to question if God chooses to reveal itself to the human being, be it in human physical form, or in a form of dreams, or in a form of voice or fire; why did God choose this Earth? Out of the whole universe and this Earth was being chosen. This could be related to quantum entanglement.

Quantum entanglement is a theory that a quantum particle that can be in two states (say particle A and B), even after being separated at a big distance, when Particle A is in the one end of the universe and Particle B is in the other end of the universe, then one of the two particles is being disturbed, it is being excited and it causes the particle to change its spin and vibration, the other pair of particle will experience changes in response to the changes imposed on the first particle. This can happen over a huge distance where there is no communication and connection between the pair of particles.

This theory can be related to God's presence in the holy book. Whatever that has been done will affect the whole universe. In other words, God does not have to repeat all the things that have been done on a planet one by one in order to reach out to living things.

So whatever God has done on Earth has its effects all over the universe and even the other multiverse.

One the ways that can we can relate multiverse and quantum entanglement is the life after death. When a person is dead he is actually at two worlds at the same time. When a person is dead, it is in a different dimension but simultaneously the dead person can still exist and dwell with the memories of his old life on Earth.

This is explained through how a particle can have its pair and how it has quantum entanglement between them. The dead person can walk around the places he used to stay and live in, he sometimes can even be perceived by those with a sixth sense. The dead person can be in the world of the living while at the same time the real realm that the person is in is at the realm of the dead.

The concept of superposition can be applied here where the dead is in the dead state and at the same time is able to view the world of the living- the dead are experiencing the world of the living and the dead at the same time.

There is a belief that when a person is dead, the soul does not go to the final destination (believed to be heaven or hell), instead, the soul will remain in the world for as long as the person is not ready to move on and attain the closure to let go of the previous life.

Some soul that is not in peace is believed to always try to communicate with the living, especially the family members to pray or to do something in order for the soul to be able to no longer have attachments to the world. Back to quantum entanglement theory, whatever happens to the soul in the living world will affect the afterlife state in which it will continue roaming around the places he used to live in and go, or transfer into the everlasting destination and meeting the creator.

We will be explaining the concept of soul in more depth in the final chapter of this book. But before going into that we still have to look at the most scientifically intriguing part of ourselves which we are dealing with every day, all day, whether we are aware of it or not, or whether we are conscious or subconscious about it.

We will be discussing the mind and how it is relatable to our faith and we will attempt to answer it in science point of view.

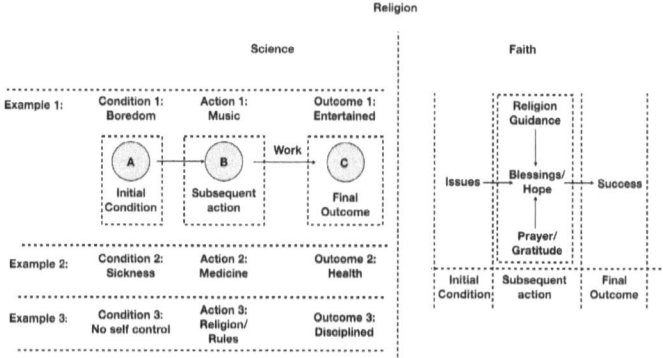

Chapter 10. Mind

The human mind is actually one example that how the universe and our life is so amazing and full of mysteries. Our human mind actually transcends dimensions. For example, the human mind can transcend time by going to the past. We could revisit what happened in the past and being given a flashback of what we could have done better or what can we repeat. Those are what we called memories.

Human's mind can also project what is in the future. We can foresee and create situations of what could possibly happen with the info that we have now, or we can create a vision with something that we created totally and originally from our mind.

The above two examples are how human mind can transcend up to 4 dimensions which are involving time: past and future; flashback and flash-forward; things are done and yet to happen. Our mind is actually a representation of how multiverse can be. Our mind can display multiple realities, of different timelines and different possibilities of outcome.
Our mind can produce something out of imagination or something that has already happened.

Our mind can think of something from the past and adding some element that didn't exist or did not happen at that time. Our mind also can visualise a future that is possible or entirely out of the context and that is without logical explanation. Our mind can think of a person from the past and use the same person to be transported to the future projection of the mind. What we have seen is stored in the memory but with the power of our brain, we can combine things that we have never seen with the things that we have seen with our own eyes. Not everything can do that.

For example, a mirror will only reflect what is projected or presented in front of them; or a normal TV monitor can only display visual view but if it has no touch-screen capabilities, it won't be able to perform another task such as recognising touch and produce output such as a pop-out message.

Sometimes our minds are so complicated that they could have their own sets of memories and characters for someone with multiple personalities.

The mind is also quite powerful and is able to repress any bad memories or trauma to save the person from that fearful event. When a person is repressing the bad memories, knowingly or unknowingly, he can only visit the memories back with a technique such as hypnosis and by meeting with a psychologist.

There are multiple theories on how human beings' are being programmed. The first idea is the Blank Slate in which it is believed human beings are born in a blank state or is like an empty vessel. They are pure and clean but will turn into whatever that is being fed and introduced to them over the course of their life especially during their formative years of childhood and adolescent.

If a person grows up in a nice, simple and happy family, that person would turn out to be an equally positive person and will have a good character and behaviour. It is depending on what is being taught and the surroundings that they were living in. In another example is if the person is growing up being exposed with a violent and angry lifestyle and grow up with an unhappy upbringing, that person will likely become problematic and having issues.

The second theory would be called the Noble Savage theory. In this idea, a person is believed to be naturally nice, selfless and peaceful. It is in the blood of a person to be benevolent, well-meaning and kind. A person is not supposed to have a bad feeling towards others or meaning to harm anyone. The term savage is to say even a person who lives and grow up in a wild who supposedly rough and scary, have good in them and naturally noble.

However, it does not mean that the mind cannot be corrupted. The corruption may occur due to fear of being overthrown and replace, competition to win in life, thirst for power and usually due to the feelings and thoughts inside and in relation with outsiders.

The third theory is the Ghost in the Shell. This concept teaches that human being is separated into two different systems. The physical state and the mind of the person. It is also theorised that human being's mind is so complex that it is a whole different study to understand it. It is also meant to say human beings can work on two different levels. A person can work if it satisfies having both physical body and the mind, or in other words, if consciousness and memories can be transferred to an empty body, the system is complete and basically is a complete living thing.

To relate all of the theories discussed with our existence, where with the Blank Slate, human beings are created as clean as an empty sheet, without the corruption from the universe. The Noble savage is the theory that supports

Blank slate, where the good in us is due to our original state which is clean, but due to the concept of survival and exposure to all sorts of ideas, the human being is exposed to corruption and the tarnishing of the clean sheet. The relationship between our faith and the concept of Ghost in a shell is better explained if we applied it in a futuristic situation: the age of digital, where computer and robots are at their peak of functionality, a human being has the computerised body and digitalised consciousness. But by the time when we managed to do so, do we still require God and religion? Heaven and hell may just be a concept or more like an outdated concept. This is when religion will be entering the higher stage as well.

As long as the human being is still mortal and limited with death, one of the ways that religion will keep teaching is the existence of Heaven and Hell. But when human being managed to be technologically advanced that death is no longer the huge part of life (except when the consciousness is corrupted and non-recoverable), religion and God will have to progress to still be applicable: as the guidance to find meaning in life.

The human being will still and always be a rational being, doing good and the right thing is not about getting yourself to be rewarded but it is about giving meaning to your life and why do you exist.

By doing good, you are creating the purpose of your existence and giving yourself the chance to contribute to the world and make the life of your loved ones to be wonderful. The human being will still realise that they have no ability to look at things beyond their time zone and they are not able to look into the future. There are still many unknowns in the life and no matter how many data, trending and information, one can only predict what will happen in the future but there are so many variables that can manipulate and change the course of the outcome.

But the most interesting thing is that whatever happened between you and God actually very much depending on the state of your mind and also what do you think of the concept of God and religion.

If you are someone who believes in God and has faith in religion, you will always find the purpose of your life: be it to glorify God, to make yourself the instrument of something bigger than yourself such as peace and charity, or to be the one to improve the life of others and your neighbours.

All of this decision is made in your mind while at the same time you are rationalising whether to believe in what you are believing in and to determine what actions and things to be done to validate the things that you believe in.

Your mind is a wonderful thing because it can think of anything with no limit. Your mind always imagines yourself to be outside of your body, where your soul is visiting at old memories, or your soul is experimenting the future scenario that your mind created. Your mind can even picture that your soul is somewhere after the death of yourself.

The mind can also do a few more stuff depending on the situation and what are they trying to achieve or what have they experienced. The main ability that the mind possessed is the ability to remember things, forgetting things, showing multiple personalities and making choices.

10.1 Remembering things.

Memories are one of the biggest element of our mind if talking about consciousness and what is in our mind, it is usually linked deep to your feelings and also memories. When we are listening to the music or songs that give us feelings, it will either link us back to old memories or assist us in creating new memories.

Especially a slow and moving music, usually it always makes us feel that it is able to transcend our senses and it goes deep into our core. Slow music is able to get us simultaneously imagining ourselves to be in a state of calmness or peace while feeling like we are in a different world or dimension. Our mind has this ability to make us feel like we are transported into different realms or teleported us into the place that we wanted to be, the place could be just a real place in real life or even imaginary destination.

Music sometimes gets us back to the memories of when we first listening to the song, or the feelings when we first discovered the music, or during a memorable event while the song is on air or any happenings associated to the song. Music and memories seem to be interrelated or at least music can bridge us to our memories- either revisiting old memories or creating new memories.

For someone with amnesia and Alzheimer, they are having problems remembering things and recollecting experiences.

Many people shared the ability to go back in time and remembering memories whenever they listen to a song, some even remember the minute details of the experience and memories. With this, it is suggested that music could be the key to gaining back memories for those who are having Amnesia and Alzheimer. Music may not necessarily be the cure for the disease but it could open a door to solving this problem.

Sometimes when you cannot eliminate something, you cope with it. A memory is sometimes a memory suppressed or a memory lost in the ocean of other memories or just the brain and mind is weak to connects one memory to another.

Maya Angelou quoted before: "They will forget what you said, but they will never forget how you made them feel". When we feel, we are not only storing memories, but our brain is working to react to that particular situation. Feelings are deeply embedded in the memory due to it is not only stored in the lobes of the brain but it is also stored in the muscle memory in which it is responsible for the motor task of your body: sometimes when a person or a situation is making you angry or excited, you react by having your brain giving signals to the heart to pump more blood and more rapidly.

Another example is when we are in a relationship with someone, one day we may no longer together with each other. Most likely you will not remember what has been said to each other and what has been done together while the relationship is still on. You may not remember what is the partner's favourite thing or the things that they loathe, you may be unsure that a particular conversation between you two and any occasion with you two in it ever happened. But what will you two remember is the feeling that you had before and the feelings that you had for each other now.

This is what we called emotional attachment. When we are emotionally attached to a particular person or something the effectiveness in remembering and storing the memories is much higher.

We tend to remember things where we have an attachment with, especially emotional attachment, this may be linked to our survivability and consciousness. When it comes to survivability, for example, a baby or a young child will only dare to be close to his or her mother or father due to the secure attachment to their parents. They will not go with strangers and will react negatively with someone in whom they don't have any emotional attachment with.   Whereas, sexuality and romantic attachment are important in order human being to copulate and populate the world and ensure the continuity of the human species.

Emotional attachment is also linked to consciousness due to our own personal liking and favouring of things- be it music choices, favourite sports and the passion towards movies or books- when a person likes a particular work, they will admire the one that is producing the work. So they will be sad when the person who has been producing the work no longer producing the same quality of work that they expect, or the person has passed away. The emotional attachment to an inanimate object such as cars and a musical instrument can be so strong that they will spend all their time and energy in ensuring their personal favourite object and belonging is in good condition.

The other ways to relate to how we remember things is by having an emotional connection.

Emotional connection is the bond that we established with someone or something that we want to keep in our life. When someone is having an emotional connection to a particular thing or someone, they will want to ensure that whatever they do will positively impact the others while the satisfaction of doing it is the reciprocation that they are expecting which is more than enough. Another example is with pet and animals, we can develop a strong bond after going through times together, for the good or the bad, and those time spent together to create the connection that we want to continue being together and close to each other in facing the world and the future.

Faith and spirituality are usually very personal and often emotionally connected to the person who is having it. Prayers are deeply personal and something you do deep in your mind and you very much will not forget the days where you experience lowliness that you seek out for help and guidance or during days where you are so happy and you were giving thanks and expressed gratitude.

These experiences connect your feelings with your consciousness and are the main reasons why when one has found their faith and spirituality, it is not easy to change their mind and to get them to give up on it. They are emotionally attached and connected with their faith.

One of the ways to get connected and express your faith is through music, music has many genres, mood and tempo. With the variation in the genre and mood, you can show different expressions such as joy, sorrow, hope, and these sometimes get you connected to your memories and at the same time gets you connected spiritually to the One you are praying to. At the same time, you are connected to the feelings related to a certain time, when your brain, your heart rate, your blood pressure and nervous system all syncing into the best condition to perform certain songs for a certain situation. The experience is something that you will always remember how it feels and sometimes you will inspire others, then the affected people will, in turn, has the personal connection to the song in which they heard from you.

10.2 Forgetting things.

How interesting is the mind when there are people who want to remember things yet there are people who want to forget things, and it can be done voluntarily or involuntarily. Not all religion is God-Centric, one of the religions that recognised the existence of God but the core belief is centralised around practice and the way of life is Buddhism.

The idea of Buddhism is in order to achieve Nirvana it is to give up on earthly worries. Earthly worries usually come from earthly things and the feeling of wanting and coveting stuff in life. Buddhism's main teaching is to forget all of the worries and the desire of wanting more. Of course, there will always be pain such as physical pain or situational pain. However, this is a temporary experience and the happenings due to us having senses and feelings.

At the end of the day when you have acknowledged that life is about attaining peace and not being held by any life's chain then we can forget the world and just focus on understanding the peaceful way of life and thinking.

This is the ability that human possesses, to control the state of mind and the ability to choose to forget on the things that don't matter. This ability is crucial not only in getting a happy life but sometimes necessary in order to cope with happenings and the unknown future.

Even in Christian teaching, it is being said to not worry about tomorrow, for tomorrow will worry itself. It is the teaching of focusing on what you have today, be grateful of your current state of living and focus your energy on something that is in front of you, rather than speculating what could or what would not happen. Many are fearing death, mostly due to too many commitments or having too much fun and power over the world that losing it is a painful and fearful prospect.

Forgetting something is also an act of survival: when we have faced with a traumatic experience, sometimes it is so bad that it can kill us from within. But a human being is a survivor, we fight or fly when facing dangers: one of the ways is by shutting down the memory and suppress it deep within the corners of the mind.

In one of the interviews by Lady Gaga when appearing in 'The Late Show with Stephen Colbert', being a sexual assault survivor, she understands how the mind works and the brain can put a traumatic experience into a box and get it shelved. This is an act where it is better to not remember the memory as those memories will cause the other part of your body to unnecessarily work hard such as giving you cold sweat or make your heart beats faster. Your mind is so powerful that it can save you from yourself and the experience could eat you up.

Sometimes, forgetting things is not a matter of choice or a matter of survivability; sometimes it is an inability to do what the brain can do. There is a condition called Aphantasia in which the person doesn't have the mind's eye, meaning the person could not visualise things that the person is reading or heard until he sees it in reality. Another mind condition is called Prosopagnosia in which the person has no ability to remember the face or in other words, the person doesn't have any facial memory or face-blind.

Sometimes, our mind is also working all the time to replace new memories and remove the data that we rarely access to in order to ensure we remember the memories that are more useful to us. But one of the mysteries of the world is that we forgot the simplest things like why are we in a particular room or shop, what conversation were we in before we got distracted, or that one thing that we wanted to do but we forgot to do, we remember the face of a person but not the name of the person.

One of the reasons may lie on how your mind works, the mind usually choose to remember something that is repetitive and the most recent ones, but what makes it sticks lies to the reason we discussed earlier: emotional connection. The impact is greater when there is a happening that brings emotional outcome to you, be it positive or negative. There are also times where the things that you wished for or prayed for happened to you in real life, most likely you will remember it for a long time due to the emotional attachment to the subject and impact that comes with it.

Another example of forgetting things are forgetting your own personality or changing to another type of you.

Multiple personalities is a good but scary example where one can totally live a different life with a completely different set of memories and ways of thinking. They can switch to a different character and behaviour. A person with multiple personalities often doesn't realise that they have such condition and they usually got amnesia or blacked out when they switched personality. It is interesting to see how the brain has the ability to totally shut down an identity and activated a new one.

At the end of the day, we need to learn how to forget and apply it in daily lives. We should forget about our lives' worries, or at least prioritise the problems to be solved first. Worrying is not going to solve anything, to have an effective and productive day, it is better to focus on identifying the root cause of a problem and find the solution. When we expand our views, forgetting things means to move on from a certain episode or frustrations.

You might have a terrible year in your job or relationship, a new year is always a good time to start over, forgetting the failures of the past and from there you can find success again.

As powerful as the mind is, and also the body, living things and human beings are not complete without another element: soul. The soul is the hardest concept to be explained through science and also the most difficult to convince on its lack of perceivable data. Although it is the hardest part to be explained, it is one of the most important parts of faith and science can offer a platform on how it came to be in our lives and what it is and what it will be across the time.

We will be entering the final chapter which is also the ultimate challenge of explaining faith and belief with the help of Science but interestingly we will need to recap and use all of the discussion idea and theory that we used across the chapters in the book to explain our soul.

Welcome to the trip of understanding soul through science!

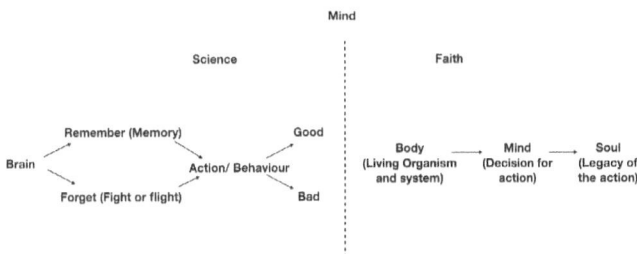

Chapter 11: The ultimate faith and science: Soul

What good will be life without purpose?

What good is life without meaning?

What good is there in life if there is nothing in us that makes us, us.

You are you and you yourself are the best version of you.

But what makes you, you?

The answer might be what is a part of you: your soul.

What is a soul?

Is it something inside of you?

Is it something that is living inside of you?

Is it the core of what you are?

When talking about soul, or rather, those who believe in the soul, believe that there is a part of you that is inside of you when you are alive and the soul will leave you once you are dead and no longer alive.

There is also a term 'soul searching', where you are going on a journey or a place in order to discover yourself or learning about yourself.

There is also a saying, you have to lose yourself to find yourself. You have to be lost to be found and that journey is your personal fight and something you have to do by yourself and for yourself.

Also, something that is not unheard of is the sentence, "good for your soul". Some of it is music or action to be done in order to make you feel good and justify your existence in this world. The soul is one of the most important parts of any religion or belief. While the soul is much of a debate when it comes to science. But what is a soul?

Back to our previous discussion about creating another you, uploading yourself into the most humanoid body ever created. Your memories can be preserved and the body can be taught the ability to feel based on enough data collected over the years to react and emphasised on a situation. But can you give a soul to the robot?

When we create a supercomputer and robot, how do you put a soul into it? Does the soul something that can be created once consciousness is achieved? Couple memories with the ability to feel and the robot is now a being with a soul?

What makes a thing alive?   Is it just by being able to breathe, pump blood, sending the electric signal from the brain to move body muscles, able to receive and process memories, the ability to remember and communicate and one is considered a soul?   Is soul something to be achieved or something that only an Intelligent Designer or Creator can create?

If in the future we do achieve technological advancement to have our memories and consciousness to be in a robot, are we but not just another object?   An object that is highly functioning, with memories and ability to communicate, interact and react- But at the end of the day just another object.   In that case, when we destroy someone or put the end to someone, what makes it considered a murder?   Killing and murder are when we robbed someone of its life, or in traditional thinking: ripping someone's soul out of the body.

So, does that mean someone with a soul is someone who is really considered a living thing?   For the past two pages alone, so many questions about soul have been pondered, and yet there are so many unanswered queries.

All of these questions probably can be answered by the scientific theory that we have presented across the book.

Let's go back to the first discussion that we have: regarding quantum bits and superposition. The soul and the body could be in a superposition: it is a part and inside of you, but at the same time your soul is a manifestation of your character and how you are with your unique thinking, style and passion in life. In other words, there's a dualism in you: your body, which is made up of a system with connected muscles and nerves; and your soul, the energy part of you that is very much unknown and does not explainable with normal science and physics.

Like superposition, the soul is a part of quantum concept where traditional physics do not apply. Your soul is the part of you that may not even possible to be measured and provide any collectable data, but it is enough to have an impact on the world and dictate how life works. The soul determines your character and your passion. Your passion is the one that dictates what you like or dislike, and the path to searching for the meaning of life and existence.

There is a saying that you can change a person's name, family, occupation, status, hairstyle... but if there is one thing that you cannot change is someone's passion: the passion for sports, passion for arts, passion for education, passion for science and technology... and the list goes on.

People change but at the core of it, they are still the same person with the same passion for something.

You can be someone who is good in science and technology, but you can also be into arts and anything that is creative or vice versa. You can be a small sized person but with a big heart (not literally), or you have a big voice. Human beings or even animals often have a distinctive behaviour or style that is not necessary following their body. Sometimes you cannot also explain intellect and athleticism.

Someone can be born of scientifically oriented parents, but somehow has zero interest with science (also he or she might be good at it) but the passion is towards artistic things such as creating music or films. You can be good at something but at the same time, you are passionate about something else. The soul is not something that you can just manufacture or create- or rather, we might never reach the capacity and ability to create a soul. We also discussed no matter how much do we know about the universe and ourselves, we only can know so much. This is due to incompleteness theorem.

No matter how much we know, we are bound by a certain limit. We can know everything within a circle, but the Intelligent Creator who created the circle would have known better and had the understanding of what happened outside the circle.

The concept of the soul may be a little hard to understand and comprehend, it may not make any sense to you at all. But as mentioned before the things you don't understand does not mean it is not correct or true. You can spend all the years trying to understand calculus or mathematics but you fail to do so, you cannot just go out and declare that it is not true and is a work of fiction. Another way of looking at the concept of soul is, just because it is still not proven by traditional physics it does not mean that it is a false theory.

To look further into how the concept of the soul can be related to the incompleteness theory, we will try to make one example. We can create a robot that is fully functioning: The robot that can move, can communicate, can react to a situation, can interact with others, can think and make a decision: a complete system. However consistent we are in developing a better robot, one iteration after another, the formula cannot solve all problem: which is to create a soul to the robot.

Sometimes, to achieve something is to not use the same approach but utilising a different way to reach it. For example, a human being is created through the biological way.

However, the most advanced robot created in the 21st century were not created out of biology, but a more organic and composite material.

Energy cannot be created or destroyed. The author believes that the soul is like an energy, it cannot be created and destroyed.

Think your body as the isolated system and your soul is like the energy. When you are alive, your soul is a form of energy that enters your body which is a form of an isolated system. At the end of the day, the soul will leave the body, the energy will leave the isolated system.
When a person is alive, the person is storing a lot of memories in the mind and at a deeper level, the data is stored up to the quantum level.

For that reason, it could be the explanation of how after a person is dead, not all is lost. The memories stay in the quantum part even though the form has converted from the earthly body into the afterlife soul.

As discussed in Chapter 3, thermodynamics law dictates that energy will never go back to the state it was before, hence the soul that is entering the afterlife stage is an entity that has memories and quantum data together with them.

A soul may be limited to the timeframe when the person is still alive (in normal time) and has to be in only one timeline, this brings us to the second discussion which is time and dimension. But when a person is experiencing out-of-body experience or is already dead, the soul is not bound by the limitation of time and possibly to be at a different dimension. The soul may even be able to visit the past, or at a different location due to its ability to be above space and time.

Another example is when a person is in trance or being possessed, the soul may not be in the body but has been going into another spacetime and dimension. To have a soul is to have something to lose, we as human have the tendency to only appreciate things that are temporary or something that will not last forever. We longed for eternal happiness, earthly wealth, a long-lasting holiday from work; but all of it is something that is only around for a while and gone after it's due.

We appreciate our own life more because we know it is temporary, we believe we will die one day no matter how advanced the technology will become. Interestingly the thing that is widely believed is that our soul will reach a permanent stage after our own death. The author believes that the soul is not something that can move out of the body only when a person is dead.

It is believed that when a person is alive, the soul is at the superposition state: where the soul is within the body of a person, but at the same time are able to be in a different location. However, when a person is dead, the soul no longer in the superposition state, however, the soul is not bound by any dimension and time, in which it entered the state where time is not in play. Hence the soul can observe the spacetime which is consisting of the past and the future.

Every religion believes that there will be a judgement day where the person is going to meet their creator and will be evaluated of their lifestyle and the deeds done on Earth. The judgement is possible due to the soul is at least of the 5th dimension and no longer bound by any time or space. The soul would be able to visit the days where a good deed was done or the days where bad decisions were made. But the author believes the soul has only the ability to observe. Only God has the ability to observe multiple inner thoughts and prayers in different places at the same time. The soul will only be able to observe one event at a time and only see what is perceivable and hear what is listenable.

In the earlier chapter, we discussed how the parallel universe is the afterlife. But in this chapter, we will also discuss the prologue to someone's life.

A soul is an entity that is always there since the creation of light. A soul could be the dark matter and when the soul is to be assimilated into the body it will be converted to light. The light is also an analogy of something that we can see which is the physical body that a normal human being can see. We also discussed in the earlier chapter that the universe is made up of dark matter. But what is dark matter? Up until now, it is not entirely sure what dark matter is and what does it contain and made of. But the author has an idea: Dark matter is the soul before it got a body.

Take an example of the bits and how the collection of 1 and 0 that make highly complex output such as the music you hear in your device and the images you see on the screen. Everything that you see on a screen such as a cartoon or any imaging is the result of millions of 1 and 0 that is in so many arrangements that result into a certain colour and a certain shape that resulting in the ultimate image that we can enjoy.

When we relate it to our universe, it is not entirely impossible that the biggest unknown part of the universe is the dark matter that is yet to be annihilated into the light. To relate to our life, the soul is the dark matter, and when it becomes a visible creature which light has shone upon them it takes a form of a body.

The whole universe is filled with particles like bits in the computer images. The ones that are dark in colour is the soul that is not visible and is beyond the normal dimension that human being can comprehend. The ones that are visible and perceivable to the eyes are the souls that have come into life and manifested in the form of living human being or even animals. But a human being is a form that is limited to time, affected by gravity, equipped with only limited capabilities. For all of the limitations known, the most powerful part of human being is the mind: because the mind's imagination is limitless and not bound by space, time, gravity and other known dimensions.

Energy cannot be created or destroyed. The dark matter made up 85% of the universe which makes it largely unknown. The dark matter is an energy that is not visible and yet to be explained. This sounds a lot like our own soul, it cannot be created or destroyed, it goes from one form to another, it cannot be observed and much of it is unexplainable. But the soul, like dark matter, is so powerful that time and gravity does not govern it. When the soul is already that powerful, imagine the Intelligent Design who created it.

The process of the superparticle broken into the 4 biggest elements in the universe, and the process of dark matter annihilated into light, are very long processes that are impossible to be observable. These are very similar to the concept of evolution. Evolution is another example of how a process of something being from one form to another can be very long.

It is very difficult to understand the process of evolution, let alone the process of quantum physics that we have to have a lot of faith and willingness to believe in: that evolution and any quantum processes are very possible but excruciating long for us to understand.

There is no such thing as nothingness in the universe: everything is something after the universe is created. Even for a superparticle (the first existing particle in the universe) to be broken into another form such as waves, light and nuclear particles, it takes years and millennials to achieve that.

The human brain and the ability that comes with it are all the product of the bigger parts of science. While the quantum scale of it is harder to decipher, such as soul and consciousness.

When a person is dead, some believes that they will be reincarnated again. These souls would have previous lives' data that will shape the person's subsequent lifestyle and character. For others, the concept of heaven and hell is believed. The author thinks that the fire of hell is not literally fire that burns, rather the fire is the analogy of the light that the soul has to receive and turn into a being again. The hell would be to relive the life and have to possess all the hardship of being human and the weaknesses that human beings usually have. While heaven is when the soul of a person who had a good life will no longer have to go through the journey of being human.

It is very interesting that we relate soul with death and God, even though not everyone believes in it but everyone knows about it and at least have put deep thoughts on it once.

God or soul should not be just something that is in our mind. When you pray, are you just connecting to the deepest part of your soul, or are you immersing yourself into the intensive side of your mind in order to reach God? There are many similarities between God and the soul. When you are trying to get connected to your soul or God, you have to allow yourself to be in an open mind, and you are opening your heart and it is not something you do subconsciously.

Our brains and bodies are wonderful, it can do a lot of things and get work done such as breathing and repairing our cells without us deciding to do it or it all happens subconsciously. But when it comes to making the connection to your soul and God, such as praying and thinking about yourself, it is a very conscious decision and it takes a lot of brain power to be able to do it. It is also not a just simple task to do, it requires guidance and direction but once you got the hold of it, you are very much in control of your perspective in life, death and yourself.

When you are using a lot of brain power, in quantum scale, more and more data and info are being stored and potentially stays long into your life and the afterlife. But why do we have to equip our soul with the good deeds that we have done? That is because we believe that there is something more in the afterlife: the body stays on Earth but the soul continues on a quantum level.

In a quantum level, the data that we have collected would be useful for us to navigate ourselves in the afterlife. Then it raises another question: why there is a need to have a life after death? The answer would be not why there is life after death, but rather our life is just a phase, a form, not created and not disturbed, it is always there and always will. Just the life that we have, the body we are in and mind that we are able to absorb knowledge are in a temporary state.

We are all a part of the universe that is scientifically driven but requires strong faith to be able to understand the things that we can compute or not.

The author believes that the soul is not merely consciousness but the soul is the highest level of consciousness. Consciousness is a state of you aware of your surroundings or the outside environment, and the higher level of consciousness is to be aware of your feelings which is the internal environment. Consciousness is a level that can be achieved but the soul is an omnipresent state but with different form across a timeline between still-born, alive, and afterlife.

Think consciousness and soul and relate it with a Gameboy. A Gameboy represents the human body and is made up of so many components and parts in order for a game to be able to be played on it. Consciousness is the level that you will only achieve if you managed to reach the higher level of the game. The sound coming out from the speaker of the Gameboy is the vibration of the sound waves. It is a representation of the soul, the wave is always there, just that when the vibration is there and the musical wave is being disturbed, then the sound and the music is heard.

The soul is always there but only manifestable when being incorporated into an earthly and humanly body.

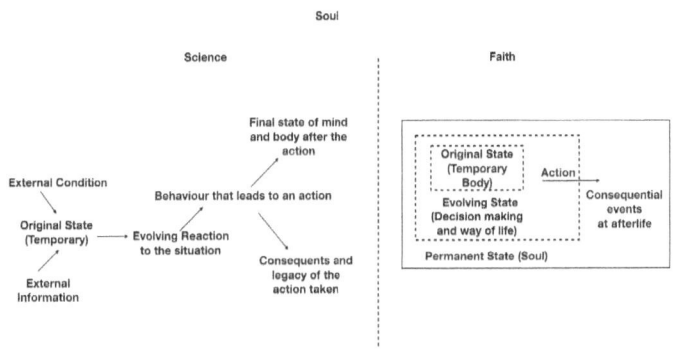

Chapter 0.1 Conclusion

The author again would like to emphasise that this book is not a religious book, nor this is a scientific journal. But the author really thinks understanding the religious faith, and the concept of God, prayers and afterlife is much more acceptable and understandable when we try to explain it with basic science.

The author really hopes for the readers that managed to follow the journey across the book and get to this part of the page, the journey feels worth it. It is with a hope that by the end of this book readers will see the world: science and faith; in a different light. The author hopes the book will spark discussions and ideas among the readers and those who follow this journey.

There are still so many things that we don't know. But maybe what we don't know is only because of the limitation that we are bound in this temporary life.

But it is a worthy effort and attempt to know as much as possible and to be able to explain and understand things as simple as possible to be relatable to all. After all, simplicity is the ultimate sophistication.

This is the end of The Science of Faith.

Important Laws, Principles, Theories, Analysis

1. The First Law of Thermodynamics: Energy cannot be created of destroyed

2. The Second Law of Thermodynamics: Entropy (Nothingness) in an isolated system will always increase

3. The Murphy's Law: Whatever that can happen, will happen

4. The Principle of Quantum Superposition: One particle can exist as two states at once

5. Mathematical Analysis: The study of changes experienced by a subject with regards to time

6. The Supersymmetry Principle: Each particle has an associated particle in other group

7. Quantum Entanglement Theory: A pair of particles continue to interact even though separated at distance

8. Theory of General Relativity: Law of Gravitation and its relation with other forces of nature (Gravity and mass)

9. Theory of Special Relativity: Elementary particles interactions and phenomena, except gravity (Space and time)

10. Simulation Theory: All reality including earth and universe is a computer simulation

11. Godel's First Incompleteness Theorem: There will always be questions that cannot be answered using the same consistent set of Axiom

12. Godel's Second Incompleteness Theorem: A system cannot demonstrate its own consistency if using different set of Axiom.
13. Bayes Theorem: Probability based on prior knowledge and conditions related to the event

References

Our Lady of Fatima and Islam
(Website Article from Return to Fatima Organization)
https://www.returntofatima.org/2015/01/our-lady-of-fatima-and-islam/

Time Entanglement Raises Quantum Mysteries
(Quanta Magazine Website news by George Musser)
https://www.quantamagazine.org/time-entanglement-raises-quantum-mysteries-20160119

Entanglement: Gravity's Long Distance Connection
(Science News Organization article by Andrew Grant)
https://www.sciencenews.org/article/entanglement-gravitys-long-distance-connection

God and Religious Toleration/The Proof of God
(Wikibooks Article)
https://en.wikibooks.org/wiki/
God_and_Religious_Toleration/
The_proof_of_God#God_and_Quantum_Physics

Relativity
(James Schombert Website Lectures)
http://abyss.uoregon.edu/~js/ast122/lectures/lec20.html

**The God Particle, Quantum Entanglement, And
The Holographic Universe**
(Business Insider Article)
https://www.businessinsider.com/the-god-particle-
quantum-entanglement-and-the-holographic-
universe-2011-4?IR=T&r=US&IR=T

**Parallel Universes, the Matrix, &
Superintelligence**
(Article by Michio Kaku in kurzweilai.net)
http://www.kurzweilai.net/parallel-universes-the-
matrix-and-superintelligence

Light
(Explain That Stuff Website article)
https://www.explainthatstuff.com/light.html

Why Consciousness Does Not Compute
(Nautilus Article by Steve Paulson on Roger Penrose
and Consciousness)
http://nautil.us/issue/47/consciousness/roger-
penrose-on-why-consciousness-does-not-compute

Scientia et Fides
(Online Paper by Sasa Horvat)
https://bib.irb.hr/datoteka/891384.13927-33008-1-
SM.pdf

Retelling the Story of Science
(Collection of lectures by Stephen M. Barr
https://www.firstthings.com/article/2003/03/
retelling-the-story-of-science

Godel's Proof
(Revised Edition by Ernest Nagel and James R.
Newman)
http://calculemus.org/cafe-aleph/raclog-arch/nagel-
newman.pdf

Gödel's Incompleteness Theorems
(Publication on Stanford Encyclopedia of
Philosophy)
https://plato.stanford.edu/entries/goedel-
incompleteness/

The Revolution of The Heavenly Spheres
(Website Book by Nicholas Copernicus)
http://bertie.ccsu.edu/naturesci/Cosmology/
Copernicus.html

What is Light?
(Article by Lambert Dolphin)
http://www.khouse.org/articles/1998/161/

Interfaith Mary
(Website articles by Ella Rozett)
http://interfaithmary.net

My Soul Is Not Me: Thomas Aquinas on Human Nature and the Afterlife
(Website Article by Adam Wood)
https://medium.com/complex-systems-channel/
the-scientific-soul-e31418ef3e5c

Quantum Leap
(QZ Website Article by Mike Murphy)
https://qz.com/924433/ibm-thinks-its-ready-to-
turn-quantum-computing-into-an-actual-business/

Superposition
(Website Article)
https://www.physicsoftheuniverse.com/
topics_quantum_superposition.html

Science and Religion
(Excerpt from Homo Deus by Yuval Noah Harari)
https://www.ynharari.com/topic/science-and-
religion/

A Ghost Story
(A Film by A24)
https://a24films.com/films/a-ghost-story

Lady Gaga Interview on The Late Show with Stephen Colbert
(Oct 5 2018 on CBS)
https://www.youtube.com/watch?v=zWaV_PTICxk

The Blank Slate

(A Book by Steven Pinker)

https://stevenpinker.com/publications/blank-slate

Supersymmetry: The Future of Physics Explained

(Article by Adam Mann on Wired Website)

https://www.wired.com/2012/07/supersymmetry-explained/

Neil deGrasse Tyson Interview on The Late Show with Stephen Colbert

(Jan 6 2018 on CBS)

https://www.youtube.com/watch?v=TgA2y-Bgi3c

God and Rev. Bayes

(Article by Victor Stenger)

https://www.csicop.org/sb/show/god_and_rev._bayes

San Junipero, Black Mirror

(A TV episode on Netflix)

https://www.imdb.com/title/tt4538072/

Lorentz Transformation

(Website Article)

http://www2.arnes.si/~gljsentvid10/ltrans.html

Inception
(A film by Legendary Pictures/Syncopy)
https://www.imdb.com/title/tt1375666/

Bloch sphere model for two-qubit pure states
(Paper by Chu-Ryang Wie
https://arxiv.org/pdf/1403.8069.pdf

Robocop
(A film by Orion Pictures)
https://www.imdb.com/title/tt1234721/

The History of Timekeeping
(Beaglesoft website article)
http://www.beaglesoft.com/maintimehistory.htm

Scout Rocket Experiment
(An Article from Georgia State University Website)
http://hyperphysics.phy-astr.gsu.edu/hbase/
Relativ/gratim.html

What is Casimir Effect
(Article by Scientific American website)
https://www.scientificamerican.com/article/what-is-
the-casimir-effec/

Vimana: Flying Machines of the Ancients
(A book by David Hatcher Childress)
https://www.amazon.com/Vimana-Machines-David-
Hatcher-Childress/dp/1939149037

Interstellar
(A film by Legendary Pictures, Syncopy and Lynda
Obst Production)
https://www.imdb.com/title/tt0816692/

Altered Carbon
(A TV Episode on Netflix)
https://www.imdb.com/title/tt2261227/

**Brian Greene Interview on The Late Show with
Stephen Colbert**
(Nov 12 2015 on CBS)
https://www.youtube.com/watch?v=0jjFjC30-4A

Lucy
(A Film by Europa Corp and Universal Pictures)
https://www.imdb.com/title/tt2872732/

**An experiment seeks to make quantum
physics visible to the naked eye**
(An experiment by University of Basel on physic.org
website
https://phys.org/news/2016-05-quantum-physics-
visible-naked-eye.html

Focus: Another Step Back for Wave-Particle Duality

(Article by Michael Schirber on APS Physics website
https://physics.aps.org/articles/v4/102

Six Feet Under

(A TV Series by HBO)
https://www.imdb.com/title/tt0248654/

Experiments for further readings

1. Lorentz Transformation (Time Dilation)
2. Beam Splitter Experiment (Duality)
3. Balls On a Sheet Experiment (Relativity)
4. Scout Rocket Experiment (Time Dilation)
5. Schrodinger's Box Experiment (Incompleteness Theorem)
6. Boiling Points of Water Experiment (Incompleteness Theorem)
7. Mind Scanning by fMRI & EEG (Wave)
8. Magnetic Resonance Imaging (Wave)
9. Casimir Effect Experiment (Superparticle/Supersymmetry)
10. Double Slit experiment (Wave Function Collapse)

www.ingramcontent.com/pod-product-compliance
Lightning Source LLC
Chambersburg PA
CBHW030615220526
45463CB00004B/1298